# Reconciling the Debate: Evidence for Caveolin-1 in Human Ventricular Cardiomyocytes

Nina

## Table of contents

expression of the ventricle myosin heavy chain-1 (VMHC1) and suppresses the atrial myosin heavy chain-1 (AMHC1).[13] The transcription factor Hey2 maintains ventricular identity by suppressing the atrial myosin light chains (Myl4 and Myl7), the SERCA2a inhibitor sarcolipin, the gap junction associated connexin-40 (Cx40) and the prohormone natriuretic peptide precursor A (Nppa).[14] Furthermore, the nuclear receptor COUP-TFII promotes the longer action potential, increased cardiomyocyte size, and development of a high density of T-tubules in ventricular cardiomyocytes.[15] Interestingly, recent mass spectrometry analysis of different human heart regions confirmed the chamber-specific proteome profile of atrial and ventricular cardiomyocytes, and among others previously discussed transcription factors and the proteins Nppa, Cx40, Myl4, and Myl7, which are differentially expressed.[16]

Directly related to the subject of this book, the proteomic profile of the human heart revealed the expression of caveolin-1 (CAV1), caveolin-2 (CAV2) and caveolin-3 (CAV3) in ventricular and atrial heart tissue by label-free mass spectrometry (SWATH-MS, Sequential Window Acquisition of All THeoretical Mass Spectra) (Table 1.1).[16] However, only CAV3 was previously accepted as isoform in heart and skeletal muscles, while CAV1 was described as predominant isoform in nonmuscle cells, for example in adipocytes and fibroblasts.[17] Accordingly, the expression of CAV1 was controversially discussed in cardiomyocytes.[18,19,20,21] However, electron microscopy studies of human ventricular cardiomyocytes[18], as well as Western blotting of cultured mouse ventricular cardiomyocytes provided evidence for CAV1 expression[19,20], while studies of isolated cardiac myocytes showed CAV1 in mouse and rat hearts.[21]

**Table 1.1 Caveolin (CAV) expression in the human heart.** SWATH-MS was used to quantify differentially expressed proteins in the left atrium (LA), right atrium (RA), left ventricle (LV) and right ventricle (RV). The post-hoc summary showed robustly detected log2 SWATH intensity areas of all three CAV isoforms across atrial and ventricular samples. Table modified from Doll et al, 2017.[16]

| Protein | LA | RA | LV | RV | Peptides | Unique peptides |
|---------|-------|-------|-------|-------|----------|-----------------|
| CAV1 | 33.04 | 32.88 | 32.56 | 32.67 | 14 | 3 |
| CAV2 | 29.23 | 28.98 | 28.75 | 28.40 | 9 | 2 |
| CAV3 | 24.61 | 25.05 | 24.46 | 25.10 | 6 | 4 |

## 1.2 Biogenesis of caveolae

CAVs are integral membrane proteins in caveolae, 50-100 nm large omega shaped nanodomain invaginations of the plasma membrane.[22] A spiked caveolar coat structure was resolved by cryo-electron microscopy, and the spikes were interpreted as cytosolic protrusions of oligomeric CAV protein complexes (Figure 1.1).[23] The oligomerization of CAVs and their interactions with cholesterol and phospholipids is thought to be essential to form the characteristic invaginated membrane shape.[23]

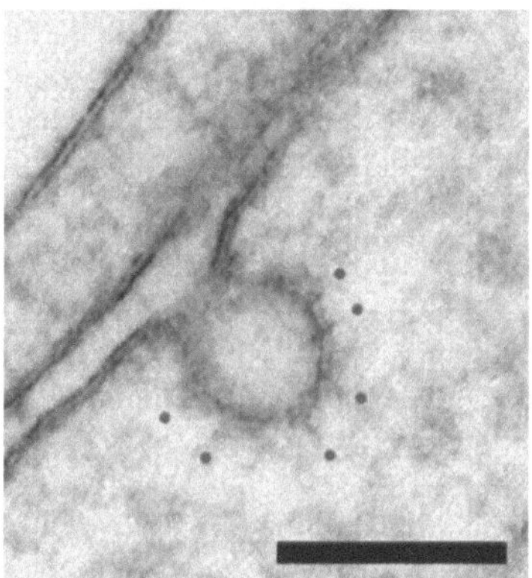

Figure 1.1 **Caveolae coat model**. Cryo-electron microscopy of an ultrathin section of a baby hamster kidney cell showing a spiked caveolar coat on the cytosolic surface; spikes are marked by red dots. Scale bar: 100 nm. Figure modified from Parton et al., 2006.[23]

CAV oligomers of 7–14 protomers are formed in the ER.[24] The CAV oligomers enter the secretory pathway to be exported to the Golgi complex in COPII vesicles. CAV oligomers associate with cholesterol in the Golgi, followed by export and assembly of oligomers in filament-like supercomplexes to form caveolae.[25] Interestingly, cholesterol is required for CAV oligomer stability, as the exit from the Golgi complex is slowed by cholesterol depletion.[25] Accordingly, Golgi exit was proposed as a step for quality control to ensure assembly of CAV multimers, which define the essential structural component of the caveolae core complex.[26,27] Caveolae are present in most mammalian cells, except for lymphocytes and hippocampal neurons.[28] Especially in mechanically active cells, such as endothelial, adipocytes and muscle cells, caveolae are highly abundant and were proposed to buffer the mechanical deformation of the plasma membrane as a protective mechanism.[29] In the heart, a high density of caveolae have been shown at the plasma membrane, increasing the surface area up to 2-fold.[22,30,31,32] Indeed, disruption of caveolar biogenesis in cardiac muscles can lead to cell damage and compensatory hypertrophy.[33] Caveolae exist as single pits with a characteristic omega-shaped structure at the cell surface,[22] while association of multiple caveolae was proposed to contribute to T-tubules biogenesis in postnatal muscle cells[5], or to multi-lobed caveolar structures called rosettes[34]. CAV3 knock-out in mice resulted in a complete loss of caveolae and decreased T-tubule density.[35] It was proposed, that the actin cytoskeleton regulates the organization of caveolae such that actin polymerization increases caveolae abundance.[36] Furthermore, the actin cytoskeleton was proposed to be involved in caveolae endocytosis and recycling.[37,38] The association of caveolae with actin filaments was documented by electron microscopy in fibroblasts[39], epithelial cells[40] and muscle cells[29].

## 1.3 CAVs are structural components of the caveolae core complex

CAVs are 21-24 kDa integral membrane proteins, which were identified as essential structural components of the caveolae core complex.[41] CAV isoforms are encoded by three genes: CAV1, CAV2, and CAV3.[22] Additionally, the mRNA of CAV1 is spliced to produce α- and β-forms, while the β-form only lacks the first 31 amino acids (Figure 1.2).[42] So far, the functional differences between α- and β-forms remain unclear.[43] The α-form was proposed as predominant CAV1 isoform[43], henceforth I refer to the α-form as CAV1 unless stated otherwise. All CAV proteins contain five functional domains: the N-terminal domain, the oligomerization domain including the scaffolding domain, the intramembrane domain, and the C-terminal domain (Figure 1.2).[44] The scaffolding domain is essential for CAV oligomerization[45], and required for caveolae biogenesis.[46] While the oligomerization, scaffolding, intramembrane and C-terminal domains are conserved across the CAV isoforms[23], the length and sequence of the N-terminal domain is highly variable.[23] Human CAV1 and CAV3 are 61 % identical.[47] CAV2 is less conserved compared to CAV1 (30%) and CAV3 (33%).[48] Both the N and C termini are predicted to be directed toward the cytoplasm, while the central intramembrane domain is predicted to form a hairpin-like structure.[44] The C-terminal domain contains three palmitoylation sites, which stabilize the membrane association of CAVs (Figure 1.2).[49,50]

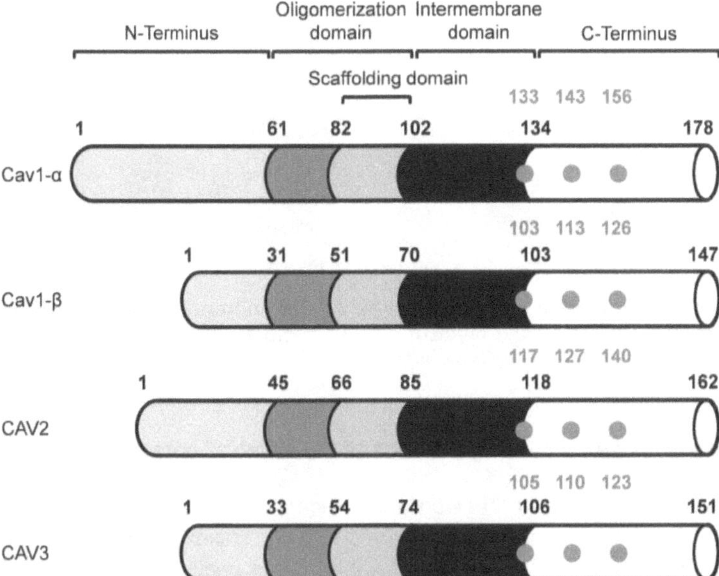


 **Domain model of human CAV isoforms.** CAV isoforms consist of five functional domains: the N-terminal domain, the oligomerization domain including the scaffolding domain, the intramembrane domain, and the C-terminal domain. The N-terminal domain is highly variable. Palmitoylation sites are indicated by green dots and residues identified by amino acid number. Figure modified from Parton et al., 2006.[23]

CAV1 and CAV3 were established as essential proteins for caveolar biogenesis[23], while CAV2 was not necessary but may contribute to the stability of CAV1-dependent caveolae invaginations.[51] CAV1 and CAV2 are co-expressed as shown for adipocytes, lung endothelial cells and fibroblasts[52] and form hetero-oligomeric complexes, which are transported to the cell surface.[53,54] In contrast, CAV3 form homo-oligomeric complexes.[48] Recently, a cryo electron microscopy (cryoEM) study revealed the first single-particle 3D structure of CAV3 oligomers in a 200 kDa nonameric disc-shaped complex with a diameter of ~165 Å.[48] It was proposed that the outer ring consists of the N-termini, while the C-termini form the central cone domain (Figure 1.3). Both, N- and C-terminal domains are pointed into the cytosol.[48] Assembly of nonameric CAV3 complexes may thus represent the building block for subsequent caveolae biogenesis.[48]

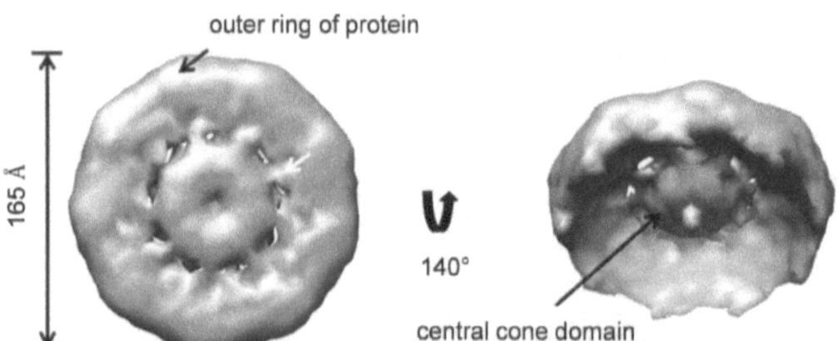


**Figure 1.3 Three-dimensional structure of the multimeric human CAV3 core complex.** Reconstituted cryoEM CAV3 structure expressed in Sf9 insect cells. Figure modified from Whiteley et al., 2012.[48]

## 1.4   Cavins are key accessory proteins of the caveolar core complex

Cavins are soluble proteins that were more recently discovered and shown to stabilize the membrane-inserted CAV1 and CAV3 complexes through protein and lipid interactions by forming an additional Cavin coat. The Cavin coat is formed via Cavin coiled-coil domain protein interactions, which drive the oligomerization in higher-ordered homo- and hetero-trimers.[55] The Cavin multimers were suggested to preassemble in the cytosol, and then bind to cholesterol- and CAV-rich membrane domains.[56] Cavins exist as four isoforms further detailed in Table 1.2. Cavin1 was shown to be an essential cytosolic coat component that can directly interact with all other Cavin isoforms.[57] Together with CAV1 or CAV3, Cavin1 was identified as essential component of the caveolar core complex.[58] Cavin1 is a ubiquitously expressed key lipid-binding protein, that directly stabilizes caveolar structures at the surface membrane.[22,57] Accordingly, Cavin1 knock-out in mice resulted in a complete loss of caveolae in skeletal muscle as evidenced by electron microscopy.[59] For in-depth discussion of Cavin functions in different cell types please refer to Kovtun, 2015.[55]

Moreover, Eps15 homology (EH) domain containing proteins and Pacsin2 (Table 1.2) were identified as components of caveolae and localized to the caveolar neck region.[60,61,62,63,64] These proteins are not essential for caveolar biogenesis, however they influence the caveolar morphology, dynamics and inhibit endocytosis of caveolae (Table 1.2).[60,61,62,63,64]

Table 1.2. **Caveolin core complex and key accessory proteins of caveolae.**
Table modified from Parton et al., 2018.[22]

| Protein | Gene | Properties |
|---|---|---|
| CAV1 | CAV1 | Integral membrane protein;[41] <br> Cholesterol binding, palmitoylated;[50] <br> Essential for caveolae biogenesis in non-muscle cells[23] |
| CAV2 | CAV2 | Integral membrane protein;[41] <br> Cholesterol binding, palmitoylated;[65] <br> Forms a complex with caveolin-1;[51] <br> Not essential for caveolae biogenesis[51] |
| CAV3 | CAV3 | Integral membrane protein;[41] <br> Cholesterol binding, palmitoylated;[49] <br> Essential for caveolae biogenesis in muscle cells[23] |
| Cavin1 | PTRF | Soluble coat protein;[55] <br> Ubiquitiniously expressed;[66] <br> Essential for caveolae biogenesis;[42] <br> Recruits Cavin2, Cavin3 and Cavin4 to caveolae[55,67] |
| Cavin2 | SDPR | Soluble coat protein;[55] <br> Only essential for caveolae biogenesis in endothelial cells;[68] <br> Oligomerizes with Cavin1[55,67] |
| Cavin3 | SRBC | Soluble coat protein;[55] <br> Not essential for caveolae biogenesis;[68] <br> Role in trafficking of caveolae;[69] <br> Oligomerizes with Cavin1[55,67] |
| Cavin4 | MURC | Soluble coat protein;[70] <br> Muscle-specific;[70] <br> Not essential for caveolae biogenesis;[71] <br> Promotes Rho/ROCK signaling;[70] <br> Oligomerizes with Cavin1[72] |
| EH domain containing proteins | EHD1 EHD2 EHD3 EHD4 | Localized to caveolae neck;[73] <br> ATPase forming ring around neck of caveolae;[60,61] <br> Inhibits endocytosis;[60,61] <br> Essential for Caveolae stabilization[60,61] |
| Pacsin2 | PACSIN2 | Localized to caveolae neck;[59] <br> Essential for caveolae stabilization;[74] <br> Recruits GTPase dynamin2, which mediates caveolae internalization by GTP-driven membrane fission[64] |

## 1.5 CAV protein interactions provide important functions

In addition to the structural function of CAVs for caveolar biogenesis[22], CAVs are thought to mediate functional protein interactions with the nitric oxide synthase (NOS) and G-protein coupled receptors (GPCRs), as well as with ion channels and ion transport proteins (Table 1.3).[75,76,77] Accordingly, human CAV mutations were shown to interfere with functionally relevant protein interactions, and lead to dysregulated mechanosensing, cell signaling or ion homeostasis (for further information, please refer to chapter 1.7). [17,78] However, a relatively large cumulative number of candidate protein interactions is based on overexpression in heterologous cell systems. Importantly, the functionality of the protein interactions was often not established under endogenous conditions of relatively low CAV expression levels (Table 1.3). For example, the proposed CAV3 interaction with the $Na^+$ channel $Na_v1.5$ was identified in HEK293 cells stably expressing $Na_v1.5$ after exogenous CAV3 overexpression.[75] Moreover, overexpression of mutant F97C-CAV3 in HEK293 cells resulted in increased late $Na^+$ currents suggesting a mechanism for action potential prolongation as the basis for the long-QT syndrome in patients.[75] However, overexpression can profoundly influence protein interactions in heterologous cell systems, as CAV1 overexpression was shown to specifically increase the pool of non-caveolar CAV1 in endosomes, but not the physiological caveolar pool.[79,80]

Table 1.3 **CAV1 and CAV3 protein interactions.** Previously published CAV1 and CAV3 protein interactions, identified by co-immunoprecipitation using heterologous overexpression or endogenous cell systems, as indicated.

| Protein | Interacting CAV isoform | Cell system |
|---|---|---|
| Src kinase (Src) | CAV1 | Overexpression in Cos7 cells[76] |
| Insulin receptor (IR) | CAV1 | Overexpression in Cos7 cells[81] |
| | CAV3 | Overexpression in HEK293 cells[82] |
| Endothelial nitric oxide synthase (eNOS) | CAV1 | aortic endothelial cells[83] |
| | CAV3 | Cardiomyocytes [84] |
| Connexin-43 (Cx43) | CAV1 | Keratinocytes from human epidermis[85] |
| | CAV3 | Mouse heart tissue[86] |
| Sodium/potassium ATPase α1 (Na/K ATPase α1) | CAV1 | Kidney tubular epithelium cells[87] |
| | CAV3 | Mouse ventricular tissue[88] |
| Potassium/sodium hyperpolarization-activated cyclic nucleotide-gated channel (HCN4) | CAV1 | Mouse SAN myocytes[89] |
| | CAV3 | Mouse SAN myocytes[89] |
| β-1-,β-2-adgeneric receptor (β-1-,β-2-AR) | CAV3 | Overexpression in HEK293 cells[90] |
| Glucose transporter (GluT4) | CAV3 | Skeletal muscle cells[91] |
| Sodium/calcium exchanger (Ncx1) | CAV3 | Mouse ventricular tissue[92] |
| Inward-rectifier potassium ion channel (Kir2.1) | CAV3 | Overexpression in Cos1 cells[93] |
| L-Type calcium channel (Ca$_V$1.2) | CAV3 | Mouse ventricular cardiomyocytes[77] |
| Sodium voltage-gated channel (Na$_V$1.5) | CAV3 | Overexpression in HEK293 cells[75] |

## 1.6 APEX2 proximity biotinylation to identify CAV3 protein networks

In this book, the engineered Ascorbate PEroXidase variant 2 from soybean (APEX2)[94,95] was used as CAV3 tag to target a proximity-based biotinylation technique for mass spectrometry analysis to the macromolecular CAV3 complex. As protein biotinylation represents a rare posttranslational modification in mammalian cells, for example mainly restricted to few mitochondrial carboxylases, exogenous biotinylation can be used to identify endogenous proteins in nanometric proximity of the APEX2 enzyme reactive cloud.[96,97] APEX2 catalyzes the oxidation of biotinphenol to a short-lived (<1 ms) biotinphenoxyl radical, a reaction which is typically activated by a 1 min $H_2O_2$ pulse.[97] The biotinphenoxyl radical reacts with electron-rich amino acids, particularly tyrosine, tryptophan, cysteine, and histidine of proteins in nanometric proximity.[97] Based on the short half-life of the biotinphenoxyl radical and electron microscopy analysis of APEX2 tagged proteins[98], a biotinylation radius of 20 nm was measured in HEK293 cells.[97] Due to the optimized high enzymatic activity of APEX2[95], the APEX2 biotinylation approach was previously shown to resolve the dynamics of protein interactions with G-protein-coupled receptors.[99] The scheme of the CAV3-tagged APEX2 biotinylation reaction is illustrated in Figure 1.4.

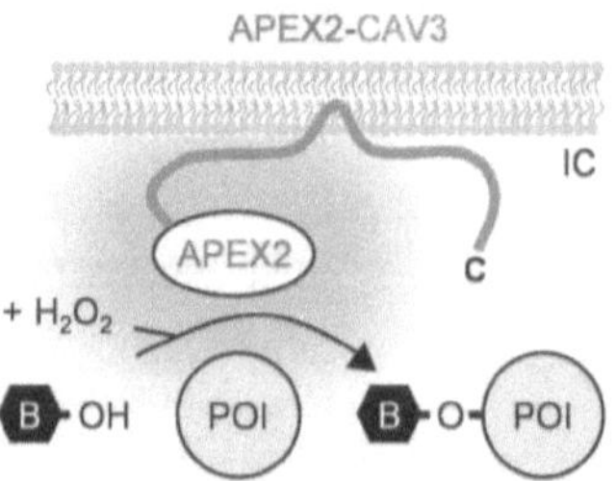


Figure 1.4 **Topological model of APEX2-CAV3 proximity-based biotinylation.** CAV3 was N-terminally tagged with APEX2. Upon cellular treatment with $H_2O_2$ for 1 min, APEX2 generates a reactive cloud of biotinphenoxyl radical molecules (red) that covalently label proteins (yellow circle) in nanometric proximity. Biotinylated proteins-of-interest (POI) are identified by mass spectrometry. B, biotinphenol; IC, intracellular

## 1.7 CAV knock-out mouse models exhibit cardiac dysfunction

While CAV1 expression in cardiomyocytes has remained controversial,[18,19,20,21] CAV1 knock-out mice were associated with a phenotype of cardiac hypertrophy.[100,101] Importantly, the hypertrophic heart changes in CAV1 knock-out mice were solely related to cardiac fibroblasts and endothelial cells.[100,101] Since CAV1 was shown to inhibit the enzymatic nitric oxide synthase (NOS) function through direct protein interactions[102], it is interesting that increased nitric oxide (NO) levels were identified in fibroblasts and endothelial cells of CAV1 knock-out mice.[100,101] Chronically increased NO levels may induce fibrosis,[103] which was proposed to cause myocardial hypertrophy.[104] Moreover, CAV1 knock-out has been shown to decrease left-ventricular conduction velocity through decreased connexin-43 expression.[105] Connexin-43 is essential for electrical coupling between myocytes at gap junctions[106]

In addition, a progressive cardiomyopathy at 4 months of age with significant hypertrophy was shown in CAV3 knock-out mice.[107] T-tubule disorganization and a decreased T-tubule density were observed by confocal microscopy in isolated ventricular cardiomyocytes from CAV3 knock-out mice.[108] Consistent with T-tubule remodeling in heart failure,[109] fewer transverse and more longitudinal tubules were documented in CAV3 knock-out cardiomyocytes.[108] Therefore, reorganization of T-tubules due to CAV3 deficiency was proposed to impair E-C coupling.[108]

## 1.8 Human CAV mutations cause a spectrum of muscle diseases

CAV1 mutations were associated with lung and vascular diseases[110], lipodystrophy[111] and breast cancer[27]. The breast cancer associated CAV1 mutation CAV1-P132L was shown to diminish caveolar biogenesis and accumulate in the Golgi through disrupted protein oligomerization in CAV1-P132L stable expressing human mammary epithelial cells.[27] Moreover, genome-wide studies have associated common CAV1 mutations with atrial fibrillation.[112,113] Human atrial tissues from patients with atrial fibrillation showed a reduced expression of CAV1, while expression of CAV2 and CAV3 was not affected.[114] However, CAV1 mutations have never been studied in striated muscles, particularly not in cardiomyocytes.

To date, 24 distinct missense mutations have been reported in the human CAV3 gene.[78] Both skeletal and heart muscle diseases have been linked to CAV3 mutations including rippling muscle disease[115], hyperCKemia[116], limb-girdle muscular dystrophy (LGMD-1C)[117] and a hypertrophic cardiomyopathy classified as long QT syndrome type 9 (LQT9)[75]. Autosomal-dominant CAV3 mutations[78] can cause a loss of CAV3 expression, as revealed by immunoblot and immunohistochemistry analysis in patient muscle biopsies.[118,119] Consistent with a severe loss of CAV3 protein expression, electron microscopy analysis of muscle samples from LGMD-1C patients showed diminished caveolae at the sarcolemma and a reorganized T-tubule system.[119] The LGMD-1C patient mutation CAV3-P104L[120] was established as a model of disrupted caveolar biogenesis.[121,117,26] Transgenic overexpression of CAV3-P104L in mice showed CAV3 protein aggregates, which were retained in the Golgi complex[121], and may induce proteasomal degradation.[117] Skeletal muscles of CAV3-P104L transgenic mice revealed proliferated ER-Golgi intermediate compartment (ERGIC) structures and abnormal localization of the dystrophin-glycoprotein complex.[26] Moreover CAV3-P104L was shown to diminish the surface expression of the insulin-dependent glucose transporter type 4 (GluT4) as well as glucose uptake in skeletal myotubes, indicating that muscle dysfunction is potentially associated with compromised transmembrane substrate uptake.[122] Related to the subject of this book, six CAV3 mutations were identified in 17 patients with the Long-QT syndrome 9, while CAV3-F97C and CAV3-S141R were proposed as disease causative CAV3 mutations.[75] Overexpression of CAV3-F97C and CAV3-S141R in HEK293 cells increased the late $Na^+$ current.[75] Mechanistically, CAV3-F97C was proposed to lose its function as NOS inhibitor,[123] while increased nitrosylation of $Na_v1.5$ channels can increase. $Na^+$ currents. [123,124] Furthermore, F97C was associated with a decreased Kir2.1 current based on co-expression in HEK293 cells.[93] Immunostaining of overexpressed CAV3-F97C in HEK293 cells co-localized CAV3-F97C with the trans-Golgi marker Golgin97, indicating CAV3-F97C accumulation in the Golgi complex.[93] Therefore, CAV3-F97C was proposed to reduce Kir2.1 channels surface expression presumably by retaining Kir2.1 channels in the Golgi complex.[93]

## 1.9 Human induced pluripotent stem cells (iPSC) derived cardiomyocytes as model of heart disease

Inducible pluripotent stem cells (iPSCs) can be generated from patient samples to study genetic diseases in the patient-specific genetic background.[125] Furthermore iPSC-derived atrial and ventricular cardiomyocytes[126] can be derived through standard protocols to study the mechanisms of cardiac diseases, as regenerative therapeutic approach, and for drug discovery.[125] Previously, the functional effects of ion channel mutations identified in long QT syndrome (LQTS) patients were studied in iPSC-derived cardiomyocytes.[127] Recently, clustered regularly interspaced short palindromic repeats (CRISPR) based on the CRISPR associated protein Cas9 has been applied for gene editing of iPSCs.[128] CRISPR/Cas9 can also be used to correct and thus prove a disease-causative mutation effect.[129] Moreover, CRISPR/Cas9 genome editing enables the expression of a targeted protein intervention under its endogenous promotor, overcoming limitations of heterologous overexpression cell systems and potentially aberrant protein interactions.[130] However, the differentiation process of the iPSC–derived cardiomyocytes is critical for phenotype development and cellular metabolism.[131,132] Long-term cultivation for 60–100 days is currently used to generate differentiated human cardiomyocytes with well-organized sarcomeric structures, functional $Ca^{2+}$ handling, and expression of cardiac-specific ion channels.[126]

## 2 Aim of this book

The aim of this book was to define the cardiac CAV protein interactions by unbiased proximity and affinity based mass spectrometry approaches. Previous studies established a large number of CAV3 protein interactions of functional relevance in heterologous cell overexpression systems. Here, we used an alternative APEX2-based ratiometric proximity assay to identify CAV3 protein interactions in living neonatal rat cardiomyocytes (NRCMs). To overcome the limitations of CAV3 overexpression, plasmid and viral vector transfection strategies were tested in parallel to express CAV3 at a similar level to endogenous CAV3. In order to validate APEX2 based candidate hits and to analyze CAV1 and CAV3 isoform-specific interactions, we used affinity proteomics and confirmed putative hits by co-immunoprecipitation. As CAV1 was recently identified both in atrial and ventricular human heart tissue by quantitative mass spectrometry methods, we investigated the relationship between CAV1 and CAV3 in adult mouse cardiomyocytes by immunoblotting and STED superresolution microscopy (nanoscopy).

Moreover, the human CAV3-P104L mutation was shown to interfere with CAV3-dependent protein interactions and to affect CAV3 oligomerization.[117] For this purpose the potential impact of the hypertrophic cardiomyopathy associated F97C and S141R mutations were analyzed by proximity proteomics, crosslinking and blue native polyacrylamidgelelektrophorese (BN-PAGE) analysis. Finally, based on the identified CAV3 interactome, the cardiomyocyte-specific disease mechanism of the monocarboxylate transporter McT1 was investigated by extracellular cell surface biotinylation. Metabolic changes were analyzed by Seahorse measurements in CRISPR/Cas9 gene-edited CAV3 knock-out and CAV3-F97C knock-in human iPSC-derived cardiomyocytes.

Major parts of my work contributed to a first author manuscript. The manuscript is part of my book in the following chapter 3, including a detailed table of my contributions. Data, which are not included in the manuscript are part of chapter 4 (**Additional Methods**) and chapter 5 (**Additional Results**).

## 3 Functional Stabilization of The Lactate-Proton Shuttle McT1 Requires Isoform-specific Caveolin Interactions Abolished by the Human CAV3-F97C Mutation

Jonas Peper, MS[1,2], Daniel Kownatzki-Danger, MS[1,2], Gunnar Weninger, PhD[1,2], David Pacheu Grau, PhD[3], Robin Hindmarsh, MS[2], Sören Brandenburg, MD[1,2,4], Tobias Kohl, PhD[1,2,4], Gerd Hasenfuss, MD[1,2,4], Michael Gotthardt, MD[5,6,7], Eva A. Rog-Zielinska, PhD[8], Bernd Wollnik, MD[4,9], Peter Rehling, PhD[3], Lukas Cyganek, PhD[2,4], Henning Urlaub, PhD[10], Christof Lenz, PhD[10], Stephan E. Lehnart, MD[1,2,4,11] ✉

[1] Cellular Biophysics and Translational Cardiology Section, Heart Research Center Göttingen, and [2] Department of Cardiology & Pneumology, University Medical Center Göttingen, Göttingen, Germany. [3] Department of Cellular Biochemistry, University Medical Center, Georg-August-University, Göttingen, Germany. [4] DZHK (German Centre for Cardiovascular Research), partner site Göttingen, Germany. [5] Neuromuscular and Cardiovascular Cell Biology, Max Delbrück Center for Molecular Medicine in the Helmholtz Association, Berlin, Germany. [6] Department of Cardiology, Virchow Klinikum, Charité - University Medicine, Berlin, Germany [7] DZHK (German Center for Cardiovascular Research), partner site Berlin, Germany. [8] University Heart Center, Faculty of Medicine, University of Freiburg, Freiburg im Breisgau, Germany. [9] Institute of Human Genetics, University Medical Center Göttingen, Göttingen, Germany. [10] Bioanalytical Mass Spectrometry, Max Planck Institute for Biophysical Chemistry, Göttingen, Germany and Bioanalytics, Institute of Clinical Chemistry, University Medical Center Göttingen, Germany.
[11] BioMET, Center for Biomedical Engineering and Technology, University of Maryland School of Medicine, Baltimore, Maryland, USA.

✉ Corresponding author: Stephan E. Lehnart, M.D.

University Medical Center Göttingen; Robert-Koch-Str. 42a, 37075 Göttingen, Germany

Tel: +49 551 39-63631; Email: slehnart@med.uni-goettingen.de

## Figure contribution

| | |
|---|---|
| Figure 3.1 | **JP** performed all experiments and data analyses |
| Figure 3.2 | **JP** performed SILAC incorporation, immunoblotting, APEX assay and data analyses; HU and CL performed mass spectrometry |
| Figure 3.3 | **JP** isolated cardiomyocytes, performed immunoblotting, immunofluorescence, co-IP, generated SWATH samples and data analyses; HU and CL performed mass spectrometry |
| Figure 3.4 | **JP** generated SWATH samples, performed co-IP, immunofluorescence and data analyses; HU and CL performed mass spectrometry |
| Figure 3.5 | **JP** performed immunoblotting, cell surface biotinylation, and bromopyruvate uptake and data analyses, **JP** and DPG performed Seahorse measurements, LC and RH differentiated and cultivated iPSC cardiomyocytes, LC performed genome editing |
| Figure 3.6 | **JP** performed all experiments and data analyses |
| Figure 3.7 | **JP** performed APEX assay and data analyses; HU and CL performed mass spectrometry |
| Figure 3.8 | **JP** performed immunoblotting, cell surface biotinylation and data analyses, **JP** and DPG performed Seahorse measurements, LC and RH differentiated and cultivated iPSC cardiomyocytes, LC performed genome editing |
| Supplement Figure 3.9 | **JP** performed all experiments and data analysis |
| Supplement Figure 3.10 | **JP** performed GO term analysis and analyzed the data, **JP** and DKD performed SILAC incorporation, HU and CL performed mass spectrometry, ERZ performed electron tomography and analyzed the data |
| Supplement Figure 3.11 | **JP** performed all experiments and data analyses |
| Supplement Figure 3.12 | **JP** performed APEX assay and data analyses; HU and CL performed mass spectrometry |
| Supplement Figure 3.13 | LC and RH cultivated iPSC, performed the experiments and data analyses, LC performed genome editing |
| Supplement Figure 3.14 | **JP** performed all experiments and data analyses |
| Supplement Figure 3.15 | **JP** performed all experiments and data analyses |
| Supplement Figure 3.16 | **JP** performed all experiments and data analyses |

**Functional Stabilization of The Lactate-Proton Shuttle McT1 Requires Isoform-specific Caveolin Interactions Abolished by the Human CAV3-F97C Mutation**

*Peper: Identification of McT1 as Caveolin3 Interactor*


Jonas Peper, MS[1,2], Daniel Kownatzki-Danger, MS[1,2], Gunnar Weninger, PhD[1,2], David Pacheu Grau, PhD[3], Robin Hindmarsh, MS[2], Sören Brandenburg, MD[1,2,4], Tobias Kohl, PhD[1,2,4], Gerd Hasenfuss, MD[1,2,4], Michael Gotthardt, MD[5,6,7], Eva A. Rog-Zielinska, PhD[8], Bernd Wollnik, MD[4,9], Peter Rehling, PhD[3], Lukas Cyganek, PhD[2,4], Henning Urlaub, PhD[10], Christof Lenz, PhD[10], Stephan E. Lehnart, MD[1,2,4,11] ✉

[1] Cellular Biophysics and Translational Cardiology Section, Heart Research Center Göttingen, and [2] Department of Cardiology & Pneumology, University Medical Center Göttingen, Göttingen, Germany. [3] Department of Cellular Biochemistry, University Medical Center, Georg-August-University, Göttingen, Germany. [4] DZHK (German Centre for Cardiovascular Research), partner site Göttingen, Germany. [5] Neuromuscular and Cardiovascular Cell Biology, Max Delbrück Center for Molecular Medicine in the Helmholtz Association, Berlin, Germany. [6] Department of Cardiology, Virchow Klinikum, Charité - University Medicine, Berlin, Germany [7] DZHK (German Center for Cardiovascular Research), partner site Berlin, Germany. [8] University Heart Center, Faculty of Medicine, University of Freiburg, Freiburg im Breisgau, Germany. [9] Institute of Human Genetics, University Medical Center Göttingen, Göttingen, Germany. [10] Bioanalytical Mass Spectrometry, Max Planck Institute for Biophysical Chemistry, Göttingen, Germany and Bioanalytics, Institute of Clinical Chemistry, University Medical Center Göttingen, Germany.
[11] BioMET, Center for Biomedical Engineering and Technology, University of Maryland School of Medicine, Baltimore, Maryland, USA.

✉ Corresponding author: Stephan E. Lehnart, M.D.

University Medical Center Göttingen; Robert-Koch-Str. 42a, 37075 Göttingen, Germany

Tel: +49 551 39-63631; Email: slehnart@med.uni-goettingen.de

## 3.1 Introduction

Cardiomyocytes exhibit a high density of caveolae in the plasma membrane increasing their surface area up to 2-fold.[1,2] Recent progress revealed how caveolae are molecularly stabilized through the coordinated actions of key lipid-binding proteins.[1] The membrane-inserted caveolin (CAV) family members CAV1 and CAV3, as well as Cavin1 were identified as essential cytosolic coat components that directly stabilize the characteristic omega-shaped caveolae structure in the surface membrane.[1,3] The caveolar neck constriction is stabilized through a ring-shaped protein complex, where Eps15 homology domain 2 (EHD2) inhibits caveolar endocytosis.[5,3]

Since CAV1 knockout mice develop heart failure and CAV1 is highly abundant in fibroblasts, it was assumed that cardiomyocytes are not primarily affected.[4] In contrast, immuno-gold EM localized CAV1 in caveolae in human cardiomyocytes,[5] and immunoblotting confirmed CAV1 expression in isolated mouse cardiomyocytes.[6] In addition, genome-wide studies have associated common CAV1 variants with cardiac conduction disease and atrial fibrillation.[7] Accordingly, a cardiomyocyte-specific mechanism of conduction disease was demonstrated in CAV1 knockout hearts.[6] However, neither the relationship between endogenous CAV1 and CAV3 nor human loss-of-function mutations were established in cardiomyocytes previously.

Both functionally important mechanoprotective and signal transduction roles have been assigned to caveolae.[1] For example, in skeletal muscle, flattening and disassembly of caveolae in response to increased stretch protects membrane integrity during muscle contraction.[2] In addition, G-protein coupled receptor and ion channel signaling have been associated with caveolae.[2,8] However, the multitude of candidate protein interactors cumulatively assigned through overexpression studies contrasts with recent observations in gene-edited cell lines showing that bulk membrane proteins are relatively depleted in caveolae.[9] Caveolar exclusion of membrane proteins was proposed to involve steric barriers provided by the coat and the ring complexes, as well as unfavorable membrane curvatures.[3,9] However, neither were putative CAV3 interactions nor CAV1 binding proteins functionally established in cardiomyocytes.[1,2]

CAV1 knock-out mouse embryonic fibroblasts with dichloroacetate shifts glucose utilization from lactate and pyruvate production to OXPHOS stimulation and cell death.[10] This was linked to mitochondrial dysfunction due to abnormal cholesterol accumulation leading to impaired mitochondrial import of the key antioxidant glutathione, such that OXPHOS increases reactive oxygen species

(ROS) sufficiently to trigger apoptosis.[10] Notably, an increase in apoptotic hepatocytes and neurons was confirmed in CAV1 knockout mice in vivo.[10] Interestingly, in the failing human heart it has been shown that lactate is an important respiratory substrate and lactate uptake through McT1 is chronically increased, apparently compensating for decreased fatty acid utilization in mitochondrial energy production.[11] Finally, during exercise lactate uptake is markedly increased in cardiomyocytes and mitochondrial lactate oxidation may then account for over 50% of oxygen consumption in the human heart.[11]

In patients, rare CAV3 mutations cause a hypertrophic cardiomyopathy classified as long-QT syndrome type 9.[8] Accordingly, overexpression of CAV3 containing the human F97C or S141R mutation in HEK293 cells stably expressing the $Na^+$ channel $Na_V1.5$ resulted in action potential prolongation.[8] On the other hand, CAV3 was found to co-purify with the dystrophin-glycoprotein complex (DGC) presumably through indirect interactions with the nitric oxide synthase,[12] whereas $Na_V1.5$ was shown to bind indirectly to the DGC through $\alpha$- and $\beta$-syntrophin.[13] Hence, the molecular nature and pathogenic mechanism of the putative CAV3 interactions each with itself in core caveolar complexes, with $Na_V1.5$, or the DGC remain unclear. As powerful live-cell as well as quantitative proteomic techniques have emerged recently, we set out to develop advanced mass spectrometry techniques to define the spectrum of cardiac CAV1 and CAV3 protein interactions in an unbiased fashion.[14,15]

Here, we identify McT1 as a member of a new class of CAV3 binding proteins and a functional link to cardiac metabolism. While this interaction does not extend to the CAV1 isoform, McT1 and CAV3 occur in functionally important membrane domains, specifically in transverse (T-) tubules. In human iPSC-cardiomyocytes CAV3 knock-out uncovered a stabilizing role both for McT1 surface expression and co-transport of small monocarboxylates, particularly for lactate/proton shuttling. Strikingly, the F97C mutation abolished the biogenesis of the multimeric CAV3 core complex and McT1 surface expression, and depressed mitochondrial respiration in human iPSC-cardiomyocytes. Hence, stabilization of functional McT1 expression in the surface membrane requires isoform-specific CAV3 protein interactions to sustain mitochondrial energy production in human iPSC-cardiomyocytes.

## 3.2  Methods

The authors declare that all supporting data including complete proteine tables are available within the article and the Supplemental methods. Detailed experimental protocols and buffer composition tables are provided in the Supplemental methods.

### 3.2.1  Ratiometric Proximity Proteomics in Live Cardiomyocytes.

We used the engineered peroxidase APEX2 to genetically tag CAV3 in living cardiomyocytes to label endogenous proteins in nanometric proximity of the macromelecular CAV3 complex via biotinylation (Figure 3.1 A). Bicistronic recombinant adenoviral vectors were generated to express V5-APEX2-CAV3 and eGFP at the lowest effective multiplicity-of-infection (MOI 1) in neonatal rat ventricular myocytes (NRCMs) for 48 h in culture. In addition, adenoviral V5-APEX2 expression served as soluble control that does not associate with CAV3 or membrane lipids. For ratiometric proteomic analysis we used stable isotope labeling by amino acids in NRCM culture and systematic label switching as 3-state SILAC approach (Figure 3.2 A). Based on 96.5% or higher SILAC incorporation in NRCMs after 13 days (Figure 3.2 B), adenoviral transfection occurred on day 11 for subsequent V5-APEX2-CAV3 or V5-APEX2 protein labeling. Biotinnylated proteins were enriched by avidin and identified by liquid chromatography-tandem-mass spectrometry (LC-MS/MS).

### 3.2.2  Label-free Sequential Window Acquisition of All THeoretical Mass Spectra (SWATH-MS)

For affinity purification (AP) followed by label-free quantification (AP-MS), CAV1 and CAV3 were immunoprecipitated from 500 µg mouse ventricular tissue. For label-free SWATH-MS quantification samples were run on 4-12 % gradient gels, cut out as a single fraction, and in-gel trypsin digested. Rabbit IgG (12-370, Merck) was used as negative control. Digested proteins were analyzed with a nanoflow chromatography system (Eksigent nanoLC425, SCIEX) hyphenated to a hybrid triple quadrupole-TOF mass spectrometer (TripleTOF 5600+, SCIEX) equipped with a Nanospray III ion source. In short, qualitative LC/MS/MS analysis was performed with a Top25 data-dependent acquisition method. For quantitative SWATH analysis, MS/MS data were acquired using 65 variable size windows14 across the 400-1,050 *m/z* range.

### 3.2.3 Blue Native (BN)-PAGE analysis, Co-immunoprecipation, and Immunoblotting

BN-PAGE was used to identify high molecular weight (MW) macromolecular complexes of the V5-APEX2-CAV3 fusion protein with endogenous CAV3 in NRCMs, as well as high MW complexes of recombinant wild-type or mutant CAV3. Reciprocal co-immunoprecipitations followed by immunoblotting were used to confirm isoform-specific CAV1 versus CAV3 protein interactions.

### 3.2.4 Human CAV3 knock-out and F97C CAV3 knock-in iPSC-cardiomyocytes

CRISPR/Cas9-mediated genome editing in human induced pluripotent stem cells (iPSCs) was applied to generate CAV3 knock-out and F97C CAV3 knock-in lines, and engineered iPSCs were directly differentiated into ventricular-like cardiomyocytes for functional analysis. The study was approved by the Ethics Committee (approval number 10/9/15) and carried out in accordance with the approved guidelines.

## 3.3  Results

### 3.3.1  Targeting the macromolecular CAV3 complex for live-cell proteomics

To label endogenous proteins in cardiomyocytes, we developed an N-terminally tagged V5-APEX2-CAV3 tool construct (Figure 3.1 A). APEX2, an engineered peroxidase, is used to biotinylate proteins in nanometric proximity (i.e., <20 nm) to CAV3 in living cells upon $H_2O_2$ treatment (Figure 3.1 A).[14,15] We hypothesized that V5-APEX2-CAV3 and endogenous CAV3 form a multimeric protein complex in neonatal rat cardiomyocytes (NRCMs) if their expression levels are similar. Using adenoviral vectors, we titrated the multiplicity of infection down to the lowest effective dose (MOI 1) and immunoblotting confirmed V5-APEX2-CAV3 expression levels similar to endogenous CAV3 (Figure 3.1 B). Since plasmid transfected NRCMs showed a significantly lower V5-APEX2-CAV3 expression (Figure 3.1 B), we used adenoviral transfection of V5-APEX2-CAV3 (MOI 1) henceforth. As CAV3 expression in Sf9 cells produced a stable disc-shaped complex,[16] we asked if V5-APEX2-CAV3 is competent to bind CAV3 in a heteromeric complex?

Firstly, we used co-immunoprecipitation followed by V5 and CAV3 immunoblotting to exclude unspecific interactions with the soluble V5-APEX2 control. Supporting our hypothesis, co-immunoprecipitation of V5-APEX2-CAV3 confirmed endogenous CAV3 as binding partner (Figure 3.1 C). Secondly, blue native (BNE) gradient gel analysis showed a high MW complex under non-denaturating conditions. CAV3 immunoblotting confirmed a major complex at ~545 kDa both in untransfected and V5-APEX2-CAV3 transfected NRCMs (Figure 3.1 D). In addition, V5 immunoblotting identified V5-APEX2-CAV3 unambiguously as exogenous component of the ~545 kDa complex (Figure 3.1 D). For functional verification, we used affinity purification mass spectrometry (AP-MS). AP-MS confirmed in living NRCMs treated for 1-min with $H_2O_2$ that V5-APEX2-CAV3 robustly labels endogenous proteins through biotinylation (Supplement Figure 3.9 B-C). Finally, confocal imaging confirmed that V5-APEX2-CAV3 co-localizes with endogenous CAV3 in NRCMs (Figure 3.1 E; see Supplement Figure 3.9 A for higher MOI doses).

### 3.3.2  Quantitative CAV3 proximity proteomics in cardiomyocytes

To develop a quantitative proteomic approach, we established a 3-state SILAC workflow for systematic label switching (Figure 3.2 A; Supplement Figure 3.10 A). NRCMs cultured in SILAC media expressing V5-APEX2-CAV3 showed a 96.5% or higher isotope incorporation (Figure 3.2 B). As ratiometric controls, we used adenoviral transfection of V5-APEX2 or eGFP based on published protocols.[15,17] V5-APEX2-CAV3 and V5-APEX2 expression was confirmed by V5 immunoblotting (Figure 3.2 C). Biotinylated proteins were enriched by affinity purification and AP-MS identified 1131 biotinylated proteins, of which 101 proteins including 9 proteins of interest (POIs) were significantly enriched for V5-APEX2-CAV3 (Figure 3.2 D; Table 8.11 (**see Appendix**)). As expected for CAV3 protein complexes, assembled in the ER and Golgi followed by vesicular trafficking to the plasma membrane, V5-APEX2-CAV3 labeling occurred in the proximity of multiple organelles, for example in ER-associated mitochondria (Supplement Figure 3.10 B).[10] Electron tomography confirmed that caveolae are situated in less than 10 nm proximity to mitochondria in cardiomyocytes (Supplement Figure 3.10 C-E), within the range of V5-APEX2-CAV3 labeling.

Based on the STRING database (v11)[18] we mapped the interaction networks of the identified POIs using the GO terms 'caveolae', 'muscle contraction', 'pyruvate metabolism', and 'iron uptake & transport' (Figure 3.2 E). Confirming our proximity proteomic strategy, all essential and muscle-specific components of the core caveolar complex, namely CAV3, Cavin1, and Cavin4 were identified consistent with earlier studies.[19,20] Surprisingly, we identified CAV1, while Pacsin2 has been associated with caveolae previously.[21] Moreover, myosin light chain (Myl2, Myl3, Myl6), actin (Acta1, Acta2 and Actc1), and troponin (Tnni1) were identified together with the Na,K-ATPase $\alpha$ and $\beta$ subunits and the Na/Ca exchanger (Ncx1) confirming earlier studies.[22,23] Finally, proteins that define key transmembrane metabolic substrate carriers were detected (Table 8.11 (**see Appendix**)). Importantly, we identified the monocarboxylate transporter 1 McT1 and the transferrin receptor 1 TfR1 as new CAV3 proximity candidates (Figure 3.2 E).

### 3.3.3  Differential CAV1 versus CAV3 expression in cardiomyocytes

As the presence of CAV1 in striated muscle cells remains controversial,[4,24] we validated CAV1 protein expression in lysates of ventricular cardiomyocytes isolated from adult mouse hearts. Immunoblotting confirmed that CAV1 is robustly expressed, visible as single band below 25 kDa, while antibody specificity was confirmed in CAV1 knock-out mouse hearts (Figure 3.3 A; Supplement Figure 3.11 A). Moreover, label-free quantification by SWATH-MS (Sequential Window Acquisition of All THeoretical Mass Spectra Mass Spectrometry) established the expression of all three mammalian CAV isoforms in isolated cardiomyocytes with, surprisingly, the highest protein area measured for CAV1 (Figure 3.3 B). SWATH-MS protein areas were previously demonstrated to correlate strongly with absolute cellular protein concentrations.[25] Out of 1816 proteins detected, the ubiquitous CAV1 isoform was ranked #205 by protein area and thus as highly abundant, whereas the muscle-specific CAV3 ranked #1105, consistent with a lower abundance (Figure 3.3 C).

To investigate the relationship between CAV1 and CAV3 in adult ventricular cardiomyocytes, we used confocal and STimulated Emission Depletion (STED) superresolution microscopy (nanoscopy). STED nanoscopy resolved CAV1 cluster signals in physiologically relevant membrane structures, namely the intercalated disk and T-tubules, but not in the lateral surface membrane (Figure 3.3 D). While the CAV1 and CAV3 signals occurred frequently adjacent to each other, co-localized signals were not observed (Figure 3.3 D). Murine CAV1 knockout myocytes confirmed specific CAV1 signals in T-tubules (Supplement Figure 3.11 B). Finally, reciprocal immunoprecipitation of CAV1 versus CAV3 in cardiomyocyte lysates indicated relatively weak heteromeric versus strong homomeric protein interactions (Figure 3.3 E). Hence, CAV1 and CAV3 clusters are differentially distributed in cardiomyocytes, presumably through their isoform-specific core complexes, but frequently juxta-positioned right next to each in T-tubules.

### 3.4.4 Isoform-specific CAV interactions

To explore the hypothesis that CAV1 and CAV3 provide macromolecular scaffolds for differential subcellular protein interactions, we analyzed mouse ventricular tissue lysates by immunoprecipitation followed by SWATH-MS. We identified 62 potential protein interactions for CAV1 and 70 for CAV3 (Supplement Figure 3.12 A-B, Table 8.14 and 8.15 (**see Appendix**)). To further dissect isoform-specific interactions, positive hits were filtered by comparing CAV3/CAV1 enrichment after permutation-based false-discovery rate analysis (p<0.05).[26] We identified each 7 interactors for CAV1 versus 23 interactors for CAV3 (Figure 3.4 A). Importantly, among POIs with an established pathophysiological role, affinity-based SWATH-MS confirmed McT1 as a previously unknown CAV3 interactor (Figure 3.4 A). Next, we compared the fold change of the logarithmic ratio each for CAV1 or CAV3 normalized to IgG (Figure 3.4 B) finding preferential CAV1 interactions with aquaporin-1, CAV1, CAV2, Cavin1 and Cavin2. In contrast, CAV3 binds preferentially to itself, the insulin-dependent glucose transporter (GluT4), to McT1, the Na,K-ATPase $\alpha$1 and $\beta$1 subunits, Connexin43, and Ncx1. Consistent with homomeric protein complexes, heteromeric interactions between CAV1 and CAV3 were not detected by affinity-based SWATH-MS.

Finally, to validate our findings in ventricular tissue lysates, we performed reciprocal immunoprecipitation experiments followed by immunoblotting. Whereas CAV1 showed an exclusive interaction with aquaporin1, we confirmed isoform-specific CAV3 interactions with connexin43, McT1, and TfR1. Again, relatively weak heteromeric interactions between CAV1 and CAV3 were apparent contrasting with strong self-interactions (Figure 3.4 C). Finally, STED nanoscopy resolved McT1 clusters as punctate signals both in the lateral surface membrane and in T-tubules of adult mouse cardiomyocytes, the former consistent with the subcellular CAV3 versus CAV1 distribution (Figure 3.4 D). As the McT1 clusters were frequently localized directly adjacent to or inside CAV3 clusters, we propose that McT1 functionally associates with CAV3 complexes in the plasma membrane. Taken together, these data confirm isoform-specific protein interactions of CAV3 with McT1 and physiologically relevant locations.

### 3.4.5 CAV3 knockout affects McT1 surface expression in human cardiomyocytes

We hypothesized that the CAV3 interaction stabilizes McT1 expression in the surface membrane. To test this, we generated a human induced pluripotent stem cell (iPSC) knock-out model (CAV3 KO) targeting the start codon of exon 1 by CRISPR/Cas9 (Supplement Figure 3.13 A). Immunoblotting of ventricular-like cardiomyocyte lysates derived from WT iPSC confirmed robust expression of CAV3 and McT1 (Figure 3.5 A). In contrast, CAV3 KO iPSC-cardiomyocytes were completely CAV3 deficient, whereas McT1 expression was decreased (Figure 3.5 A). To explore if the McT1 decrease is functionally relevant in the sarcolemma, we applied extracellular surface biotinylation to living cardiomyocytes in culture. Surface biotinylated proteins were enriched by pull-down and McT1 identified by immunoblotting. Indeed, McT1 was specifically decreased in the surface membrane of CAV3 KO cardiomyocytes (Figure 3.5 B).

To investigate if decreased McT1 surface expression functionally impacts substrate uptake, we exposed human iPSC-cardiomyocytes to extracellular 3-bromopyruvate (3-BP), a glycolysis-disrupting compound previously established as McT1-specific substrate.[27,28] Accordingly, WT and CAV3 KO cardiomyocytes were treated with 3-BP (50 µM) and cell viability assessed by lactate dehydrogenase (LDH) release using published protocols.[29] Consistent with McT1 loss in the surface membrane leading to decreased 3-BP uptake, LDH release was significantly decreased in human CAV3 KO cardiomyocytes (Figure 3.5 C).

McT1 is thought to represent the major pathway for lactate uptake in the heart,[30] and is upregulated in heart failure.[11] To assess if human CAV3 KO cardiomyocytes experience substrate-dependent metabolic limitations, we measured oxygen consumption and extracellular acidification using Seahorse protocols established for iPSCs previously.[31] While oxygen consumption was normal in CAV3 KO cardiomyocytes (Figure 3.5 D), extracellular acidification was significantly decreased (Figure 3.5 E). Moreover, inhibiting ATP synthesis with oligomycin increased extracellular acidification maximally in WT but not in CAV3 KO cardiomyocytes consistent with impaired proton-coupled lactate export (Figure 3.5 E). Finally, mitochondrial uncoupling by ionophore (FCCP) or electron transport chain inhibition (Antimycin+Rotenone) did not result in further changes. Together, these experiments established that CAV3 KO destabilizes functional McT1 expression and extracellular acidification through the lactate/proton shuttle in human cardiomyocytes.

### 3.4.6 The human F97C but not the S141R mutation affects CAV3 oligomerization

Based on the cardiomyopathy allele frequency cut-off provided by ExAC,[32] only F97C and S141R were confirmed as potentially pathogenic CAV3 variants (Supplement Table 3.1). In cardiomyocytes, however, it is unknown if and how F97C or S141R affect CAV3 protein interactions. For proximity proteomic analysis, we transfected NRCMs with mutant V5-APEX2-CAV3-F97C or V5-APEX2-CAV3-S141R adenoviral vectors (MOI 1). While endogenous CAV3 expression was not changed, we found a decreased level of V5-APEX2-CAV3-F97C relative to the V5-APEX2-CAV3-S141R or WT V5-APEX2-CAV3 (Supplement Figure 3.14 A). Confocal imaging revealed that V5-APEX2-CAV3-F97C accumulated in a perinuclear fashion (Supplement Figure 3.14 B). In contrast, V5-APEX2-CAV3-S141R was distributed similar to endogenous CAV3 (Supplement Figure 3.14 B). To identify the nature of the perinuclear V5-APEX2-CAV3-F97C accumulation, co-localization with the trans-Golgi marker P115 previously established in cardiomyocytes indicated a trafficking problem (Supplement Figure 3.14 C).[33] Hence, extending earlier findings from heterologous model systems to cardiomyocytes,[34] the F97C mutation is predicted to disrupt the biogenesis of trafficking-competent hetero-oligomeric complexes with endogenous CAV3.

To assess the mutation-specific impact on the biogenesis of the CAV3 complex, firstly we used co-immunoprecipitation of NRCM lysates followed by CAV3 and V5 immunoblotting. This showed that both and V5-APEX2-CAV3-F97C and V5-APEX2-CAV3-S141R can bind to endogenous CAV3 (Figure 3.6 A). Moreover, as negative control we showed that soluble V5-APEX2 does not bind to endogenous CAV3 (Figure 3.6 A). Secondly, to analyze F97C and S141R depedent V5-APEX2-CAV3 oligomerization in the absence of endogenous CAV3, we used overexpression in HEK293 cells to analyze homomeric protein complexes under native conditions in BNE gels. While we confirmed a high MW complex each for the mutant V5-APEX2-CAV3-S141R and WT, the F97C mutation diminished the biogenesis of the major ~545 kDa complex (Figure 3.6 B).

Finally, to capture smaller oligomeric assemblies, we treated living HEK293 cells transfected with WT or V5-APEX2-CAV3-F97C with the cross-linker DSS. CAV3 immunoblotting revealed dimers, trimers, and higher oligomers for WT V5-APEX2-CAV3, with dimers and trimers increasing in a dose-dependent fashion (Figure 3.6 C). Strikingly, the F97C mutation disrupted each oligomeric state (Figure 3.6 C). Together, these data established that the F97C but not the S141R mutation affects the biogenesis of V5-APEX2-CAV3 complexes with endogenous CAV3 in cardiomyocytes.

### 3.4.7  Proximity proteomic analysis of CAV3 mutations

Used ratiometric proximity proteomics based on 95% or higher SILAC incorporation (Supplement Figure 3.15 A), we explored the impact of CAV3 mutations in the NRCM model. Robust biotinylation of endogenous proteins by V5-APEX2-CAV3-F97C was confirmed in living NRCMs (Supplement Figure 3.15 B). AP-MS detected 986 biotinylated proteins, 64 of which were enriched including the muscle-specific CAV3 and Cavin4 (Figure 3.7 A). However, the physiologically relevant proximity with Cavin1, McT1, and NCX1 was diminished (Figure 3.7 A). Furthermore, the number of mitochondrial and plasma membrane proteins in proximity to V5-APEX2-CAV3-F97C was substantially decreased, whereas the proximity with Golgi-associated proteins was strongly increased (Supplement Figure 3.15 C; Table 8.12 (**see Appendix**)).

Using STRING[18] based GO term analysis, we compared WT V5-APEX2-CAV3 and V5-APEX2-CAV3-F97C enriched proteins. This showed that the F97C mutation diminshed the proximity in protein networks relevant for pyruvate utilization including McT1 and the mitochondrial respiratory chain (Supplement Figure 3.15 D, Table 8.12 (**see Appendix**)). Moreover, F97C caused an increased proximity with proteins involved in 'COPI-dependent Golgi to ER retrograde trafficking' and 'COPII mediated vesicle transport' (Supplement Figure 3.15 E). To confirm that the F97C mutation disrupts the core caveolar complex, we compared the logarithmic ratio of V5-APEX2-CAV3-F97C versus WT V5-APEX2-CAV3 over V5-APEX2. F97C strongly increased the ratio consistent with Golgi accumulation (Figure 3.7 B). In addition, the Cavin1 and Cavin4 ratios were significantly decreased (Figure 3.7 B) confirming a loss of proximity with key cytosolic proteins necessary for caveolae stabilization.

In contrast, when we repeated the same analysis for the S141R mutation the proximity with Cavin1 was preserved (Figure 3.7 C), consistent with our finding of a normal subcellular distribution of the V5-APEX2-CAV3-S141R protein in NRCMs. Nonetheless, the proximity with McT1, Ncx1, and TfR1 was also not detected for S141R (Figure 3.7 C) similar to the F97C mutation. When we compared the logarithmic ratio of V5-APEX2-CAV3-S141R versus WT V5-APEX2-CAV3 over V5-APEX2, we found no significant differences indicative of Golgi accumulation (Figure 3.7 D). Together, these data indicate that the CAV3 mutation F97C profoundly impact the assembly of the caveolar core complex and post-Golgi trafficking.

### 3.4.8 The F97C mutation disrupts McT1-depedent human cardiomyocyte functions

Since the F97C mutation disrupts McT1 expression in the surface membrane of NRCMs, we hypothesized that transmembrane transport of small monocarboxylate substrates is impaired by F97C in human cardiomyocytes. To test this, we generated a human iPSC knock-in model (F97C KI) via CRISPR/Cas9 inserting the mutation in exon 2 of the human CAV3 gene (Supplement Figure 3.13 C). We analyzed lysates of F97C KI iPSC-cardiomyocytes by immunoblotting and showed that CAV3-F97C and McT1 expression were similar to the WT control (Figure 3.8 A). Strikingly, extracellular surface biotinylation of F97C KI cardiomyocytes revealed a 97% loss in McT1 surface expression (Figure 3.8 B). As expected for the negative control, β-Actin was not labeled by extracellular surface biotinylation (Figure 3.8 B).

To assess if the CAV3-F97C mutation disrupts substrate-dependent energy metabolism in human cardiomyocytes, we used Seahorse measurements. Importantly, both extracellular acidification and oxygen consumption were constitutively depressed in F97C KI cardiomyocytes at baseline (Figure 3.8 C-D). Oligomycin maximally increased extracellular acidification in WT but not in F97C KI cardiomyocytes (Figure 3.8 C) while oxygen consumption remained significantly more decreased in F97C KI cardiomyocytes (Figure 3.8 D). Importantly, oxygen consumption remained significantly depressed after FCCP treatment only in F97C KI cardiomyocytes (Figure 3.8 D). Finally, inhibition of electron transport by antimycin A and rotenone decreased oxygen consumption more in F97C KI than in WT cardiomyocytes (Figure 3.8 D). We calculated the glycolytic ATP production rate through mitochondrial oxidative phosphorylation using published protocols,[35] confirming that F97C KI diminished mitochondrial ATP production (Supplement Figure 3.16 A-B). Taken together, destabilizing functional McT1 surface expression the CAV3-F97C mutation constitutively affected proton/lactate export and mitochondrial respiration, which established a key molecular role of the isoform-specific McT1 protein interaction as a disease target and a new human heart disease model.

## 3.5 Discussion

Combining the strengths of proximity- and affinity-based proteomics, we have characterized previously unknown, isoform-specific CAV1 versus CAV3 protein interactions in cardiomyocytes. Reciprocal co-immunoprecipitation experiments confirmed McT1 and TfR1 as new CAV3-specific binding partners of immediate relevance for cardiac metabolism, while aquaporin1 was identified as CAV1-specific interactor. Interestingly, quantitative proteomics identified CAV1 as highly abundant isoform contrasting with less muscle-specific CAV3 protein in cardiomyocytes. Consistent with a functionally important and CAV3-dependent McT1 interaction in the surface membrane, CAV3 knock-out uncovered a stabilizing role for transmembrane proton/lactate shuttling in human cardiomyocytes. Moreover, the CAV3-F97C mutation suspended the oligomerization of the core caveolar complex and abrogated McT1 surface expression in human cardiomyocytes. Importantly, while genome editing established a human cardiomyocyte model with low physiological levels of the CAV3-F97C mutant protein, McT1 destabilization with depression of mitochondrial respiration defines a new molecular and metabolic framework of cardiomyopathy mechanisms.

STED nanoscopy showed that CAV3 and McT1 frequently occur in immediate proximity to each other in the lateral cardiomyocyte surface membrane and in T-tubules. Immunogold EM studies demonstrated previously that McT1 is highly expressed in caveolae, intercalated disks, and T-tubules, but only the latter were located adjacent to mitochondria.[36] Here we discovered a new role of CAV3 as an isoform-specific interactor of McT1. CRISPR/Cas9 knock-out in human cardiomyocytes established a causal role, since McT1 expression was specifically decreased in the plasma membrane of CAV3 deficient cells. McT1 is known to facilitate the proton-coupled transport of small monocarboxylates, most importantly of lactate and pyruvate.[30] During exercise lactate represents a major cardiac energy source that may account for over 50% of oxygen consumption.[37] Whereas cardiac ischemia drives lactate efflux from affected cells,[38] chronic heart failure leads to significantly increased McT1 protein expression and lactate uptake.[11] Given that McT1 has a prominent role both in physiological and pathological cardiac stress adaptation, our discovery that CAV3 interacts with McT1 and functionally stabilizes substrate metabolism in human cardiomyocytes is highly relevant.

As cultured iPSC-derived cardiomyocytes predominantly utilize glucose for ATP production,[39] energy homeostasis depends directly on McT1 surface expression and proton-coupled metabolite export.[30] Consistent with this model both CAV3 knock-out and CAV3-F97C knock-in resulted in decreased extracellular acidification. However, only cardiomyocytes expressing the CAV3-F97C showed a complete loss of McT1 surface expression. Interestingly, pharmacologic inhibition of monocarboxylate transport in tumor cells rapidly increases intracellular lactate, whereas ATP and glutathione (GSH) synthesis are decreased, contributing to mitochondrial dysfunction.[40] In analogy, human cardiomyoctes expressing the CAV3-F97C mutant protein exhibited a severe McT1 loss-of-function with decreased mitochondrial respiration and ATP production.

In addition, the F97C mutation resulted in a loss of proximity with Cavin1, Pacsin2, Ehd3 and Ehd4, indicating that a disrupted biogenesis of the CAV3 core complex prevents spatial associations with key proteins involved in caveolae biogenesis.[1] Interestingly, Golgi accumulation and defective mechanosensing were recently demonstrated for the CAV3 mutations P28L and R26Q in skeletal myofibers of patient biopsies.[41] As expected from confocal imaging studies showing perinuclear V5-APEX2-CAV3-F97C but not V5-APEX2-CAV3-S141R accumulation, the F97C mutation increased the proximity with Golgi-associated proteins. Notably, both the F97C and S141R mutation resulted in unphysiological proteomic proximities with 26S proteasome subunits (Psmd1, Psmd3, and Psmd14) and polyubiquitin C that were not detected for WT V5-APEX2-CAV3, which implicates that the unfolded protein response was activated in NRCMs. Together with the human iPSC-cardiomyocyte data these shows the broader impact of the CAV3 mutation F97C.

Quantitative proteomics and superresolution imaging established that CAV1 is highly abundant in the T-tubules in adult mouse cardiomyocytes. Our data thus overcome the prevailing notion that CAV1 is expressed predominantly in non-muscle cells.[1] In addition, immunolabeling freeze-fracture EM studies localized CAV1 in caveolae of human heart sections previously.[5] Here, STED nanoscopy showed that CAV1 and CAV3 are not co-localized, although clusters of the two isoforms occurred frequently immediately adjacent to each other. This led to the hypothesis of isoform-specific subcellular nanodomain functions based on unique protein interactions of CAV1 versus CAV3. In line with our data, the CAV3 mutation P104L associated with limb girdle muscular dystrophy was shown to diminish insulin-induced surface expression of GluT4 and glucose uptake in skeletal myotubes.[42] Finally, directly corresponding with the P104L mutation, the CAV1 mutation P132L was associated with extracardiac pathologies. Taken together, our findings of isoform-specific protein interactions provide an important template for future studies to explore the molecular impact of human CAV1 mutations in the context of cardiac muscle function.

CAV1 knockout in mice has been shown to decrease left-ventricular conduction velocity through decreased connexin-43 expression.[6] As STED imaging showed CAV1 signals both in WT and CAV1 KO mouse cardiomyocytes at the intercalated disc, we have to assume unspecific antibody binding. Nonetheless, we confirmed robust CAV1 expression in mouse cardiomyocytes by immunoblotting in WT versus CAV1 KO mouse cardiomyocytes and quantitatively by SWATH-MS in mouse ventricles. Importantly, our proteomic analysis showed that the CAV3 interaction with connexin43 and the $Na^+/Ca^{2+}$ exchanger are isoform-specific, whereas aquaporin1 was confirmed as CAV1-specific interactor through co-immunoprecipitation experiments. While human CAV1 and CAV3 are 61 % identical, only CAV1 exhibits an extended N-terminal domain subject to Src phosphorylation at tyrosine-14, augmenting Src binding to CAV1.[43] As the CAV1 variant P132L represents a well-established model of disrupted caveolae biogenesis,[44] follow-up studies will need to explore the impact of this particular human mutation on aquaporin1 function in cardiomyocytes.

Excessive overexpression in heterologous cell system has been shown to interfere with caveolae biogenesis, to lead to aberrant CAV1 trafficking, and to increase the pool of non-caveolar CAV1.[44,45] For example, 4 hours after CAV1 transfection in CV1 fibroblasts most of the overexpressed CAV1 failed to co-localize with endogenous CAV1 in caveolae, instead accumulating in the late endosome.[46] Therefore, we have carefully titrated adenoviral expression of V5-APEX2-CAV3 to the lowest effective level and showed this was similar to endogenous CAV3, which resulted in preserved CAV3 trafficking. In sharp contrast, at an MOI of 3 or higher V5-APEX2-CAV3 accumulated in Golgi organelles in NRCMs.

Genome editing of NIH3T3 cells to enable CAV1 expression at low endogenous levels demonstrated recently that caveolae are endocytosed at a very low rate and that bulk membrane proteins are excluded from caveolae.[9] Therefore, we used genome editing to establish low physiological levels of the CAV3-F97C in human cardiomyocytes and in line with our quantitative analysis identifying lower CAV3 than CAV1 levels in mouse cardiomyocytes. In contrast, earlier studies used overexpression of CAV3-F97C in HEK293 cells to infer a Nav1.5 channel interaction as cause of the long-QT syndrome.[8] Of note, unbiased affinity- and proximity-based proteomic analysis did not detect Nav1.5 as CAV1 or CAV3 interactor in ventricular cardiomyocytes. Consistent with our findings, a 3-fold transgenic overexpression of WT CAV3 causes a degenerative cardiomyopathy in mice and diminished dystrophin expression.[47] Hence we reason that our novel gene-edited human CAV3-F97C KI iPSC model has overcome significant limitations associated with heterologous overexpression systems.

In summary, we have developed a proximity proteomic technique that identified McT1 as putative CAV3 interactor in cardiomyocytes. As a second isoform-specific approach, affinity proteomics was used, which established CAV1- versus CAV3-specific interactors. For CAV3 we identified GluT4, McT1 and TfR1 as a new class of isoform-specific interactors relevant for cardiac energy metabolism, whereas aquaporin1 was identified for CAV1. Hence, combining proximity and affinity proteomics, we demonstrate that previously unknown interactors of CAV complexes can be detected with high specificity, providing a comprehensive strategy for systematic functional analysis. Furthermore, we show that McT1, an abundant cardiac lactate/proton co-transporter, requires CAV3 for functional surface expression in human cardiomyocytes. In contrast, the CAV3-F97C mutation disrupted the biogenesis of caveolar core complexes and destabilized McT1-dependent substrate-transport and mitochondrial function in human cardiomyocytes. These observations highlight the potential of in situ protein labeling to screen for new components in macromolecular complexes in the physiological context of cardiac cells. Characterization of novel interactors of CAV1 and CAV3 complexes is central to understand isoform-specific functions, cardiac cell biology, disease mechanisms, and to develop new therapeutic rationales for example to stabilize McT1 and cardiac metabolism during increased cardiac stress.

## 3.6  Acknowledgments

We are grateful for excellent technical assistance by Birgit Schumann, Brigitte Korff, Laura Cyganek, Lisa Neuenroth, Thierry Wasselin, Nadine Gotzmann, Martina Grohe, Yvonne Hintz, Lisa Krebs and Yvonne Wedekind; to Carolin Wichmann for high-pressure freezing and sample preparation for electron tomography; and to Martin Schorb from the Electron Microscopy core facility at the EMBL Heidelberg. This project was supported by Deutsche Forschungsgemeinschaft through the SFB1002 to LC (project S01), to SEL (projects A09 and S02), to PR (A06), to GH (project D01); and through the SFB1190 to SEL (project P03) and HU (project Z02). SB received financial support through the clinician scientist program "Translational Medicine" of the University Medical Center Göttingen. EARZ is funded by a Emmy Noether Fellowship by Deutsche Forschungsgemeinschaft (#396913060).

## 3.7  Conflict of interest

The authors have declared no conflict of interest.

## 3.8  Author contributions

JP, CL, and SEL designed the studies. JP, DKD, GW, DPG, RH, SB, TK, MH, ERZ, BW, HU, LC, CL, and SEL performed the research and analyzed the data. H.U. supervised MS analyses. GH provided expertise about human heart samples and cardiac energetics. PR and DPG provided expertise about mitochondrial metabolism. JP, CL, and SEL wrote the manuscript, and all authors contributed to the final version.

## 3.9 References

1. Parton RG. Caveolae: Structure, Function, and Relationship to Disease. *Annu Rev Cell Dev Biol*. 2018;34:111–136.

2. Sinha B, Köster D, Ruez R, Gonnord P, Bastiani M, Abankwa D, Stan RV, Butler-Browne G, Vedie B, Johannes L, Morone N, Parton RG, Raposo G, Sens P, Lamaze C, Nassoy P. Cells Respond to Mechanical Stress by Rapid Disassembly of Caveolae. *Cell*. 2011;144:402–413.

3. Ludwig A, Howard G, Mendoza-Topaz C, Deerinck T, Mackey M, Sandin S, Ellisman MH, Nichols BJ. Molecular Composition and Ultrastructure of the Caveolar Coat Complex. *PLoS Biol*. 2013;11:e1001640.

4. Cohen AW, Park DS, Woodman SE, Williams TM, Chandra M, Shirani J, Pereira de Souza A, Kitsis RN, Russell RG, Weiss LM, Tang B, Jelicks LA, Factor SM, Shtutin V, Tanowitz HB, Lisanti MP. Caveolin-1 null mice develop cardiac hypertrophy with hyperactivation of p42/44 MAP kinase in cardiac fibroblasts. *American Journal of Physiology-Cell Physiology*. 2003;284:C457–C474.

5. Robenek H, Weissen-Plenz G, Severs NJ. Freeze-fracture replica immunolabelling reveals caveolin-1 in the human cardiomyocyte plasma membrane. *Journal of Cellular and Molecular Medicine*. 2008;12:2519–2521.

6. Yang K-C, Rutledge CA, Mao M, Bakhshi FR, Xie A, Liu H, Bonini MG, Patel HH, Minshall RD, Dudley SC. Caveolin-1 Modulates Cardiac Gap Junction Homeostasis and Arrhythmogenecity by Regulating cSrc Tyrosine Kinase. *Circ Arrhythm Electrophysiol*. 2014;7:701–710.

7. Ellinor PT, Lunetta KL, Albert CM, Glazer NL, Ritchie MD, Smith AV, Arking DE, Müller-Nurasyid M, Krijthe BP, Lubitz SA, Bis JC, Chung MK, Dörr M, Ozaki K, Roberts JD, Smith JG, Pfeufer A, Sinner MF, Lohman K, Ding J, Smith NL, Smith JD, Rienstra M, Rice KM, Van Wagoner DR, Magnani JW, Wakili R, Clauss S, Rotter JI, Steinbeck G, Launer LJ, Davies RW, Borkovich M, Harris TB, Lin H, Völker U, Völzke H, Milan DJ, Hofman A, Boerwinkle E, Chen LY, Soliman EZ, Voight BF, Li G, Chakravarti A, Kubo M, Tedrow UB, Rose LM, Ridker PM, Conen D, Tsunoda T, Furukawa T, Sotoodehnia N, Xu S, Kamatani N, Levy D, Nakamura Y, Parvez B, Mahida S, Furie KL, Rosand J, Muhammad R, Psaty BM, Meitinger T, Perz S, Wichmann H-E, Witteman JCM, Kao WHL, Kathiresan S, Roden DM, Uitterlinden AG, Rivadeneira F, McKnight B, Sjögren M, Newman AB, Liu Y, Gollob MH, Melander O, Tanaka T, Stricker BHC, Felix SB, Alonso A, Darbar D, Barnard J, Chasman DI, Heckbert SR, Benjamin EJ, Gudnason V, Kääb S. Meta-analysis identifies six new susceptibility loci for atrial fibrillation. *Nat Genet*. 2012;44:670–675.

8. Vatta M, Ackerman MJ, Ye B, Makielski JC, Ughanze EE, Taylor EW, Tester DJ, Balijepalli RC, Foell JD, Li Z, Kamp TJ, Towbin JA. Mutant Caveolin-3

Induces Persistent Late Sodium Current and Is Associated With Long-QT Syndrome. *Circulation*. 2006;114:2104–2112.

9. Shvets E, Bitsikas V, Howard G, Hansen CG, Nichols BJ. Dynamic caveolae exclude bulk membrane proteins and are required for sorting of excess glycosphingolipids. *Nat Commun*. 2015;6:6867.

10. Bosch M, Marí M, Herms A, Fernández A, Fajardo A, Kassan A, Giralt A, Colell A, Balgoma D, Barbero E, González-Moreno E, Matias N, Tebar F, Balsinde J, Camps M, Enrich C, Gross SP, García-Ruiz C, Pérez-Navarro E, Fernández-Checa JC, Pol A. Caveolin-1 Deficiency Causes Cholesterol-Dependent Mitochondrial Dysfunction and Apoptotic Susceptibility. *Current Biology*. 2011;21:681–686.

11. Jóhannsson E, Lunde PK, Heddle C, Sjaastad I, Thomas MJ, Bergersen L, Halestrap AP, Blackstad TW, Ottersen OP, Sejersted OM. Upregulation of the Cardiac Monocarboxylate Transporter MCT1 in a Rat Model of Congestive Heart Failure. *Circulation*. 2001;104:729–734.

12. McNally E, de Sa Moreira E, Duggan DJ, Bonnemann CG, Lisanti MP, Lidov HG, Vainzof M, Passos-Bueno MR, Hoffman EP, Zatz M, Kunkel LM. Caveolin-3 in muscular dystrophy. *Human Molecular Genetics*. 1998;7:871–877.

13. Gavillet B, Rougier J-S, Domenighetti AA, Behar R, Boixel C, Ruchat P, Lehr H-A, Pedrazzini T, Abriel H. Cardiac Sodium Channel Na $_v$ 1.5 Is Regulated by a Multiprotein Complex Composed of Syntrophins and Dystrophin. *Circulation Research*. 2006;99:407–414.

14. Lam SS, Martell JD, Kamer KJ, Deerinck TJ, Ellisman MH, Mootha VK, Ting AY. Directed evolution of APEX2 for electron microscopy and proximity labeling. *Nat Methods*. 2015;12:51–54.

15. Hung V, Zou P, Rhee H-W, Udeshi ND, Cracan V, Svinkina T, Carr SA, Mootha VK, Ting AY. Proteomic Mapping of the Human Mitochondrial Intermembrane Space in Live Cells via Ratiometric APEX Tagging. *Molecular Cell*. 2014;55:332–341.

16. Whiteley G, Collins RF, Kitmitto A. Characterization of the Molecular Architecture of Human Caveolin-3 and Interaction with the Skeletal Muscle Ryanodine Receptor. *J Biol Chem*. 2012;287:40302–40316.

17. Ong S-E, Blagoev B, Kratchmarova I, Kristensen DB, Steen H, Pandey A, Mann M. Stable Isotope Labeling by Amino Acids in Cell Culture, SILAC, as a Simple and Accurate Approach to Expression Proteomics. *Mol Cell Proteomics*. 2002;1:376–386.

18. Szklarczyk D, Gable AL, Lyon D, Junge A, Wyder S, Huerta-Cepas J, Simonovic M, Doncheva NT, Morris JH, Bork P, Jensen LJ, Mering C von.

STRING v11: protein–protein association networks with increased coverage, supporting functional discovery in genome-wide experimental datasets. *Nucleic Acids Research*. 2019;47:D607–D613.

19. Parton RG, Collins BM. Unraveling the architecture of caveolae. *Proc Natl Acad Sci USA*. 2016;113:14170–14172.

20. Bastiani M, Parton RG. Caveolae at a glance. *J Cell Sci*. 2010;123:3831–3836.

21. Hansen CG, Howard G, Nichols BJ. Pacsin 2 is recruited to caveolae and functions in caveolar biogenesis. *Journal of Cell Science*. 2011;124:2777–2785.

22. Liu L, Askari A. β-Subunit of cardiac $Na^+$-$K^+$-ATPase dictates the concentration of the functional enzyme in caveolae. *American Journal of Physiology-Cell Physiology*. 2006;291:C569–C578.

23. Bossuyt J, Taylor BE, James-Kracke M, Hale CC. Evidence for cardiac sodium-calcium exchanger association with caveolin-3. *FEBS Letters*. 2002;511:113–117.

24. Volonte D, McTiernan CF, Drab M, Kasper M, Galbiati F. Caveolin-1 and caveolin-3 form heterooligomeric complexes in atrial cardiac myocytes that are required for doxorubicin-induced apoptosis. *American Journal of Physiology-Heart and Circulatory Physiology*. 2008;294:H392–H401.

25. Schubert OT, Ludwig C, Kogadeeva M, Zimmermann M, Rosenberger G, Gengenbacher M, Gillet LC, Collins BC, Röst HL, Kaufmann SHE, Sauer U, Aebersold R. Absolute Proteome Composition and Dynamics during Dormancy and Resuscitation of Mycobacterium tuberculosis. *Cell Host Microbe*. 2015;18:96–108.

26. Tusher VG, Tibshirani R, Chu G. Significance analysis of microarrays applied to the ionizing radiation response. *Proc Natl Acad Sci USA*. 2001;98:5116–5121.

27. Birsoy K, Wang T, Possemato R, Yilmaz OH, Koch CE, Chen WW, Hutchins AW, Gultekin Y, Peterson TR, Carette JE, Brummelkamp TR, Clish CB, Sabatını DM. MCT1-mediated transport of a toxic molecule is an effective strategy for targeting glycolytic tumors. *Nat Genet*. 2013;45:104–108.

28. Sprowl-Tanio S, Habowski AN, Pate KT, McQuade MM, Wang K, Edwards RA, Grun F, Lyou Y, Waterman ML. Lactate/pyruvate transporter MCT-1 is a direct Wnt target that confers sensitivity to 3-bromopyruvate in colon cancer. *Cancer Metab*. 2016;4:20.

29. Ehrke E, Arend C, Dringen R. 3-bromopyruvate inhibits glycolysis, depletes cellular glutathione, and compromises the viability of cultured primary rat

astrocytes: 3-BP and Astrocytes. *Journal of Neuroscience Research*. 2015;93:1138–1146.

30. Garcia CK, Goldstein JL, Pathak RK, Anderson RG, Brown MS. Molecular characterization of a membrane transporter for lactate, pyruvate, and other monocarboxylates: implications for the Cori cycle. *Cell*. 1994;76:865–873.

31. Rose S, Frye RE, Slattery J, Wynne R, Tippett M, Pavliv O, Melnyk S, James SJ. Oxidative Stress Induces Mitochondrial Dysfunction in a Subset of Autism Lymphoblastoid Cell Lines in a Well-Matched Case Control Cohort. *PLoS ONE*. 2014;9:e85436.

32. Whiffin N, Minikel E, Walsh R, O'Donnell-Luria AH, Karczewski K, Ing AY, Barton PJR, Funke B, Cook SA, MacArthur D, Ware JS. Using high-resolution variant frequencies to empower clinical genome interpretation. *Genet Med*. 2017;19:1151–1158.

33. Arakel EC, Brandenburg S, Uchida K, Zhang H, Lin Y-W, Kohl T, Schrul B, Sulkin MS, Efimov IR, Nichols CG, Lehnart SE, Schwappach B. Tuning the electrical properties of the heart by differential trafficking of KATP ion channel complexes. *Journal of Cell Science*. 2014;127:2106–2119.

34. Vaidyanathan R, Vega AL, Song C, Zhou Q, Tan B, Berger S, Makielski JC, Eckhardt LL. The Interaction of Caveolin 3 Protein with the Potassium Inward Rectifier Channel Kir2.1. *J Biol Chem*. 2013;288:17472–17480.

35. Mookerjee SA, Gerencser AA, Nicholls DG, Brand MD. Quantifying intracellular rates of glycolytic and oxidative ATP production and consumption using extracellular flux measurements. *J Biol Chem*. 2018;293:12649–12652.

36. Jóhannsson E, Nagelhus EA, McCullagh KJ, Sejersted OM, Blackstad TW, Bonen A, Ottersen OP. Cellular and subcellular expression of the monocarboxylate transporter MCT1 in rat heart. A high-resolution immunogold analysis. *Circ Res*. 1997;80:400–407.

37. van der Vusse GJ, de Groot MJ. Interrelationship between lactate and cardiac fatty acid metabolism. *Mol Cell Biochem*. 1992;116:11–17.

38. Vandenberg JI, Metcalfe JC, Grace AA. Mechanisms of pHi recovery after global ischemia in the perfused heart. *Circ Res*. 1993;72:993–1003.

39. Rana P, Anson B, Engle S, Will Y. Characterization of Human-Induced Pluripotent Stem Cell–Derived Cardiomyocytes: Bioenergetics and Utilization in Safety Screening. *Toxicol Sci*. 2012;130:117–131.

40. Doherty JR, Yang C, Scott KEN, Cameron MD, Fallahi M, Li W, Hall MA, Amelio AL, Mishra JK, Li F, Tortosa M, Genau HM, Rounbehler RJ, Lu Y, Dang ChiV, Kumar KG, Butler AA, Bannister TD, Hooper AT, Unsal-Kacmaz K, Roush WR, Cleveland JL. Blocking Lactate Export by Inhibiting the Myc

Target MCT1 Disables Glycolysis and Glutathione Synthesis. *Cancer Res.* 2014;74:908–920.

41. Dewulf M, Köster DV, Sinha B, Lesegno CV de, Chambon V, Bigot A, Bensalah M, Negroni E, Tardif N, Podkalicka J, Johannes L, Nassoy P, Butler-Browne G, Lamaze C, Blouin CM. Dystrophy-associated caveolin-3 mutations reveal that caveolae couple IL6/STAT3 signaling with mechanosensing in human muscle cells. *Nat Commun.* 2019;10:1–13.

42. Deng YF, Huang YY, Lu WS, Huang YH, Xian J, Wei HQ, Huang Q. The Caveolin-3 P104L mutation of LGMD-1C leads to disordered glucose metabolism in muscle cells. *Biochem Biophys Res Commun.* 2017;486:218–223.

43. Cao H, Courchesne WE, Mastick CC. A Phosphotyrosine-dependent Protein Interaction Screen Reveals a Role for Phosphorylation of Caveolin-1 on Tyrosine 14 RECRUITMENT OF C-TERMINAL Src KINASE. *J Biol Chem.* 2002;277:8771–8774.

44. Han B, Copeland CA, Tiwari A, Kenworthy AK. Assembly and Turnover of Caveolae: What Do We Really Know? *Front Cell Dev Biol.* 2016;4.

45. Parton RG, del Pozo MA. Caveolae as plasma membrane sensors, protectors and organizers. *Nat Rev Mol Cell Biol.* 2013;14:98–112.

46. Hayer A, Stoeber M, Ritz D, Engel S, Meyer HH, Helenius A. Caveolin-1 is ubiquitinated and targeted to intralumenal vesicles in endolysosomes for degradation. *J Cell Biol.* 2010;191:615–629.

47. Aravamudan B, Volonte D, Ramani R, Gursoy E, Lisanti MP, London B, Galbiati F. Transgenic overexpression of caveolin-3 in the heart induces a cardiomyopathic phenotype. *Hum Mol Genet.* 2003;12:2777–2788.

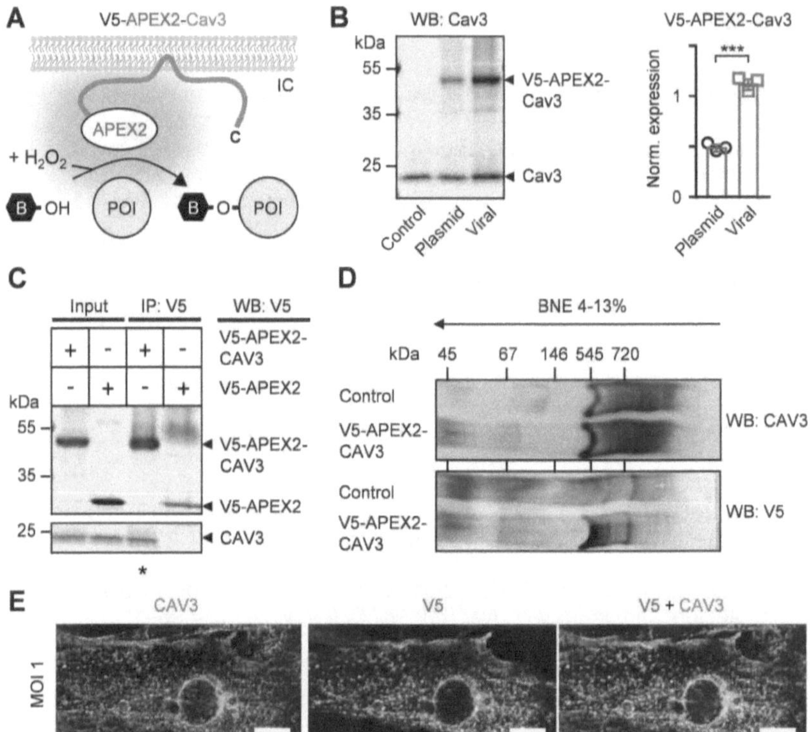



Figure 3.1 **Targeting the CAV3 complex for proximity proteomics. A,** CAV3 was N-terminally tagged by V5 and APEX2 for expression in neonatal rat cardiomyocytes (NRCM). Upon treatment of NRCMs with peroxide (1 mM $H_2O_2$ for 1 min) APEX2 generates a reactive cloud of biotinphenoxyl radical (red) that covalently labels proteins-of-interest (POI) in nanometric proximity. B, biotinphenol; IC, intracellular **B,** Representative immunoblot comparing V5-APEX2-CAV3 expressed by plasmid versus adenoviral transfection in NRCMs. Bar graph summarizing V5-APEX2-CAV3 expression normalized to endogenous CAV3. n=3, *** p<0.001, Student's t-test. **C,** V5 co-immunoprecipitation (IP) followed by immunoblotting confirmed that V5-APEX2-CAV3 binds to endogenous CAV3 (⋆) but not the soluble V5-APEX2 control in NRCMs. n=3. **D,** Blue native (BNE) gradient gel separation and immunoblotting of V5-APEX2-CAV3 in NRCMs. A major high MW complex was detected by CAV3 immunoblotting (≥545 kDa) in untransfected (control) and V5-APEX2-CAV3 transfected NRCMs. V5 immunoblotting confirmed V5-APEX2-CAV3 in the high MW complex in adenovirally transfected but not in untransfected NRCMs (control). n=3. **E,** Confocal imaging of adenovirally transfected NRCM (MOI 1) showing co-localized V5-APEX2-CAV3 with endogenous CAV3. Scale bars 10 µm.

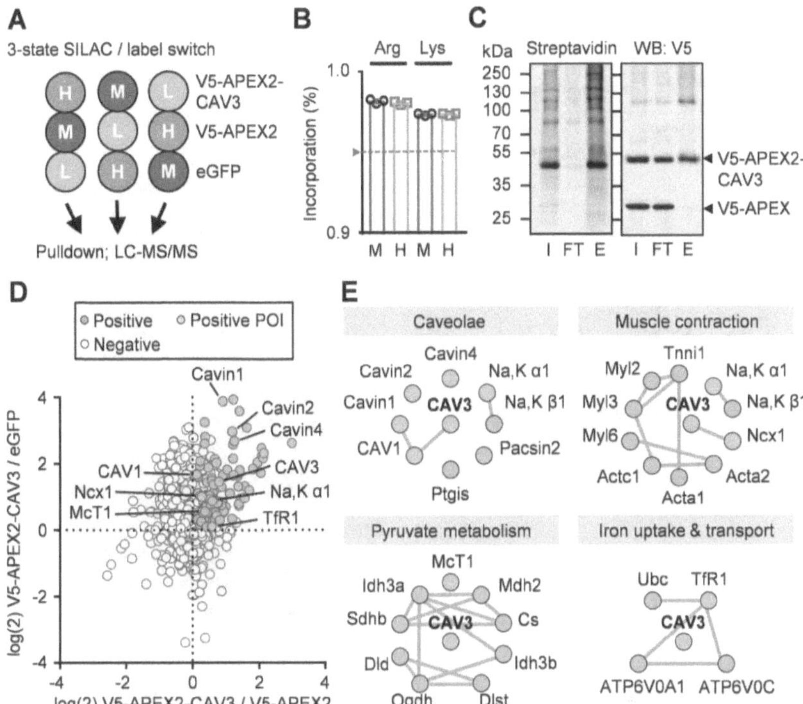

Figure 3.2 **Ratiometric proximity proteomics identifies new CAV3 interactors. A,** Systematic workflow for quantitative NRCM protein labeling by 3-state SILAC switching in 35 mm culture dishes. NRCM were labeled with light (L), medium (M), or heavy (H) L-arginine and L-lysine isotopes as indicated, followed by adenoviral transfection of V5-APEX2-CAV3 versus V5-APEX2 (control-1) or eGFP (control-2). **B,** LC-MS/MS analysis showed >96.5% L-arginine (Arg) and L-lysine (Lys) incorporation (red line, 95%). n=3. **C,** APEX2 biotinylated proteins were captured by avidin. I, input; FT, flow through; E, eluate. V5-APEX2-CAV3 and V5-APEX2 expression confirmed by V5 immunoblotting. n=3. **D,** Scatter plot based on the indicated logarithmic ratios of enriched proteins identified by LC-MS/MS. Positive hits are represented by blue and yellow color (p<0.05; z-test), the latter highlighting functionally relevant hits identified by GO analysis. n=3. **E,** Exploration of CAV3 interactions based on the STRING database for the GO terms CAVeolae, muscle contraction, pyruvate metabolism, and iron uptake. Coloring analogous to panel **D**. Protein-protein interactions are represented by grey lines based on a confidence score >0.7.

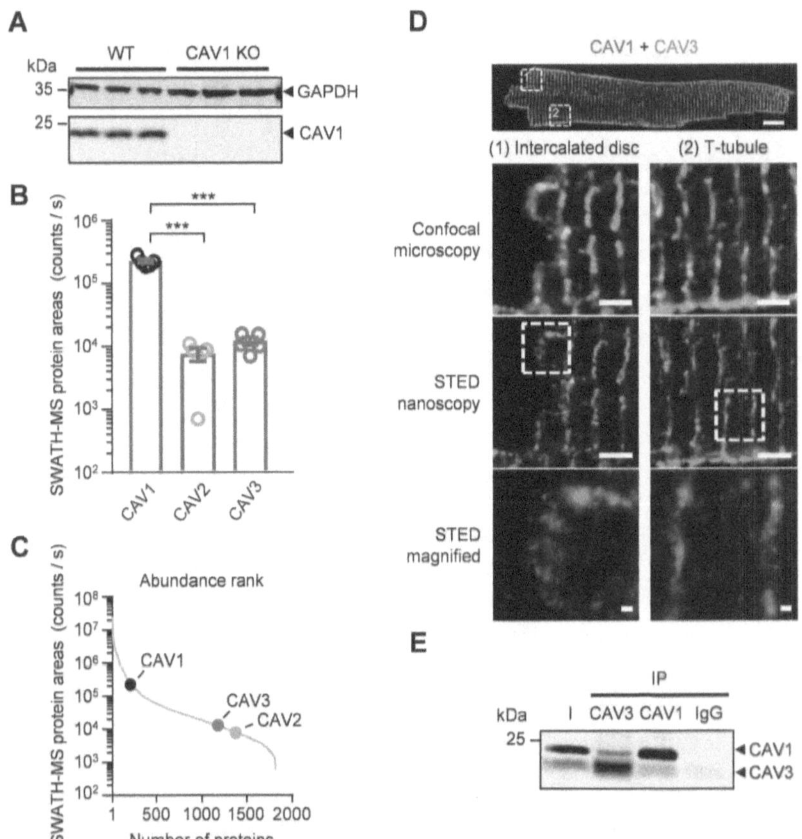



**Figure 3.3 CAV1 is differentially distributed in ventricular cardiomyocytes.** **A**, Immunoblotting showed a specific CAV1 signal in ventricular cardiomyocytes of wild-type hearts, confirmed in CAV1 knockout mouse hearts. n=3. **B**, SWATH-MS was used to estimate the relative concentration of the caveolin isoforms. CAV1 protein area was significantly higher compared to CAV2 or CAV3. n=5. *** p<0.001, ANOVA. **C**, Ranking of all proteins detected by SWATH-MS protein area. CAV1 ranks among the most abundant proteins. n=5. **D**, Confocal and STED co-immunofluorescence imaging of CAV1 and CAV3 clusters in ventricular myocytes. Dashed boxes indicate magnified regions of interest at the intercalated disk and transverse tubules, where STED superresolution nanoscopy resolved differential CAV1 versus CAV3 cluster distributions. Scale bars: top 10 µm; Confocal microscopy 2 µm; STED nanoscopy 2 µm; STED magnified 200 nm. **E**, Reciprocal co-immunoprecipitation of CAV1 and CAV3. Immunoblotting indicating strong homomeric versus weak heteromeric interactions between native CAV1/CAV3. Negative control, rabbit IgG. n=3.

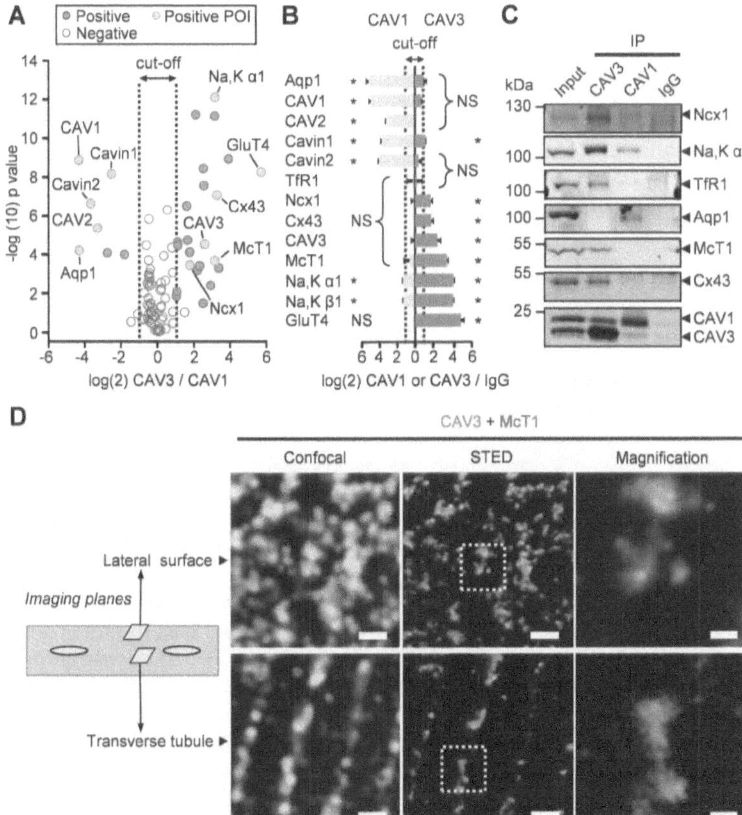


**Figure 3.4 Identification of differential protein interactions by AP-MS.** **A,** Volcano plot comparing CAV1 and CAV3 interacting proteins identified by AP-MS. Logarithmic ratios identify enriched CAV1 and CAV3 interacting proteins as indicated by positive hits (blue circles) or functionally relevant proteins of interest (yellow circles). Positive hits and proteins of interest were identified by permutation-based false-discovery rate analysis (t-test, p>0.05, FDR=5%, S0=0.1) and logarithmic cut-off >1 (dashed line). Negative hits were excluded based on the same criteria. n=3. **B,** Bar graph comparing the logarithmic ratio (control IgG) of candidate protein interactions between CAV1 and/or CAV3. A log(2) fold-change >1 was used as cut-off (dashed lanes). * p<0.001, Student's t-test. **C,** Immunoprecipitation followed by immunoblotting was used to confirm candidate protein interactions. Rabbit IgG served as negative control. n=3. **D,** Confocal and STED co-immunofluorescence imaging of CAV3 and McT1 in ventricular myocytes. The cartoon of a ventricular cardiomyocyte corresponds with the subcellular imaging planes subjected to confocal and STED imaging. Dashed boxes indicate high-power magnifications. Scale bars: image panels 1 µm; magnifications 200 nm.

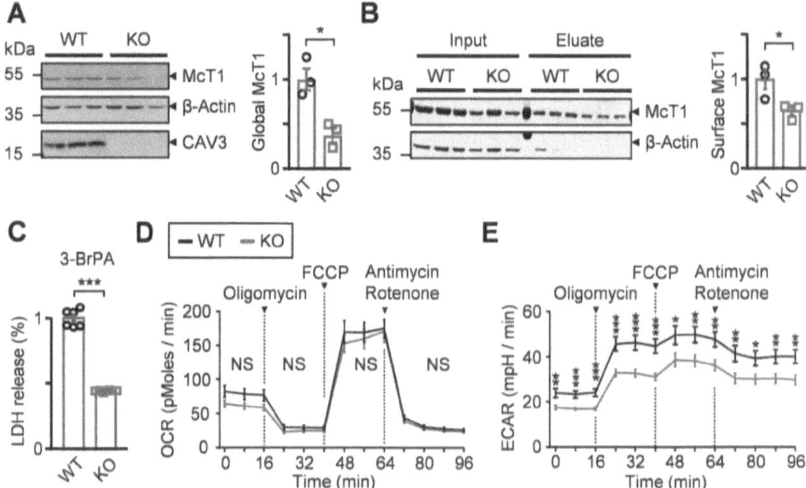

Figure 3.5 **CAV3 KO disrupts McT1 function in human cardiomyocytes. A**, Immunoblot analysis of human iPSC-derived CAV3 knockout (KO) cardiomyocytes. CAV3 and Mct1 were robustly expressed, while CAV3 signals were confirmed by using CAV3 KO cardiomyocytes. Bar graph showing significant reduction of global Mct1 expression in CAV3 KO cardiomyocytes normalized to β-Actin. n=3. Student's t-test, * p<0.05. **B**, Extracellular protein biotinylation was used in living human cardiomyocytes to assess Mct1 surface expression. Biotinylated proteins were enriched by pull-down and Mct1 identified by immunoblotting in the eluated fraction. Vice versa, β-Actin immunoblotting was used as negative cytosolic protein labeling control. Bar graph showing a significant loss of surface Mct1 in CAV3 KO versus WT cardiomyocytes. n=3. Student's t-test, * p<0.05. **C**, Uptake of 3-bromopyruvate (3-BrPA), a glycolysis-disrupting compound, was determined by cell viability based on extracellular release of lactate dehydrogenase (LDH). Bar graph showing a significant reduced LDH release for CAV3 KO cardiomyocytes incubated with 50 µM 3-BrPA. n=6. Student's t-test, *** p<0.001. **D**, The oxygen consumption rate (OCR) of human cardiomyocytes was not affected by CAV3 KO. n=32. Student's t-test. **E**, The extracellular acidification rate (ECAR) was blunted by CAV3 KO at baseline, which was exaggerated by oligomycin treatment. n=32. Student's t-test, * p<0.05; **p<0.01; *** p<0.001.

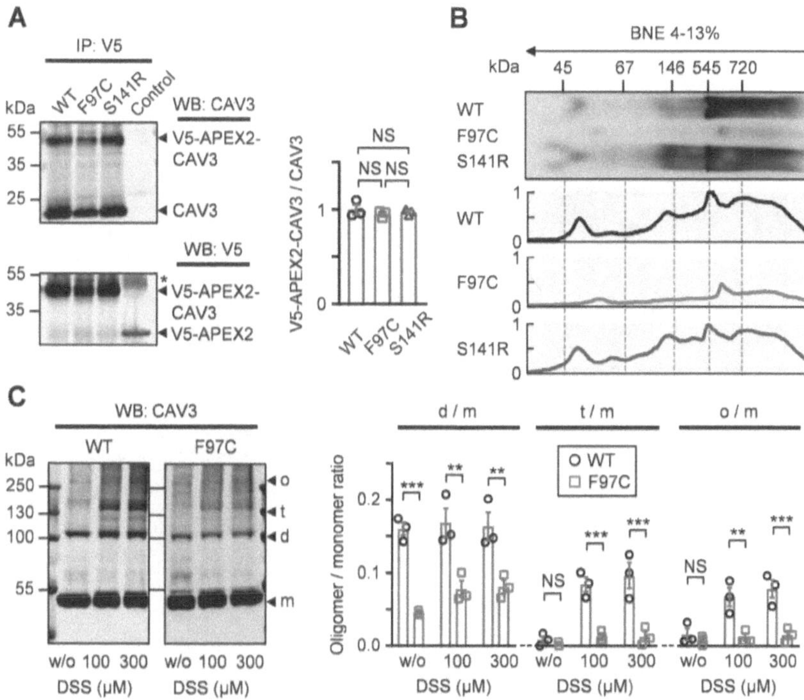

Figure 3.6 **Proteomic targeting of the human CAV3 variants F97C and S141R**. **A**, Reciprocal co-immunoprecipitation of NRCM lysates and immunoblotting showed that recombinant F97C and S141R and endogenous CAV3 form stable protein complexes. V5-APEX2, negative control. *, IgG band. Bar graph excluding significant differences in the recombinant over native expression ratio. ANOVA, n=3. **B**, BNE gradient gel separation followed by CAV3 immunoblotting of WT, F97C and S141R expression in the absence of native CAV3 in HEK293 cells. The major oligomeric WT complex (≥545 kDa) was detected for S141R but abolished by the F97C mutation. Intensity line profiles document the loss of the major signal peak for F97C. n=3. **C**, Live-cell cross-linking with di-succinimidyl-suberat (DSS) revealed significant differences between WT and F97C oligomerization in HEK293 cells. Control without DSS treatment (w/o). CAV3 immunoblotting showed monomeric (m), dimeric (d), trimeric (t), and higher oligomeric (o) products. Bar graphs document the disruption of oligomerization, normalized to the monomeric form, by F97C relative to WT. Student's t-test ** p<0.01, *** p<0.001.

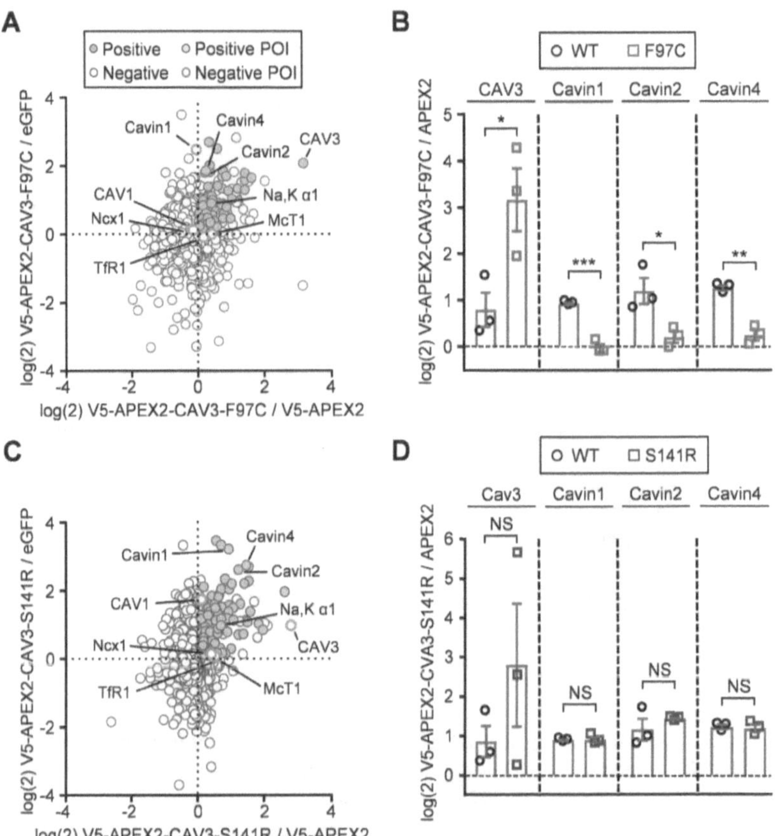

Figure 3.7 **CAV3 mutations disrupting physiological protein proximities. A,** Ratiometric proximity proteomic analysis of NRCMs expressing V5-APEX2-CAV3-F97C. Logarithmic ratio showing enriched positive versus negative hits. Cut-off criteria (logarithmic ratio <0) are indicated by dashed lines. Positive hits and POIs (p<0.05, z-test) are indicated by filled circles. Red circles indicate negative POIs. n=3. **B,** Bar graphs comparing the logarithmic ratio of essential components of the caveolar complex between WT and V5-APEX2-CAV3-F97C. n=3. Student's t-test * p<0.05, ** p<0.01, *** p<0.001. **C,** Proximity proteomic analysis of proteins labeled by V5-APEX2-CAV3-S141R. Legend same as in A. **D,** Bar graphs comparing the logarithmic ratio of the indicated caveolar complex proteins labeled by WT or V5-APEX2-CAV3-S141R. n=3.

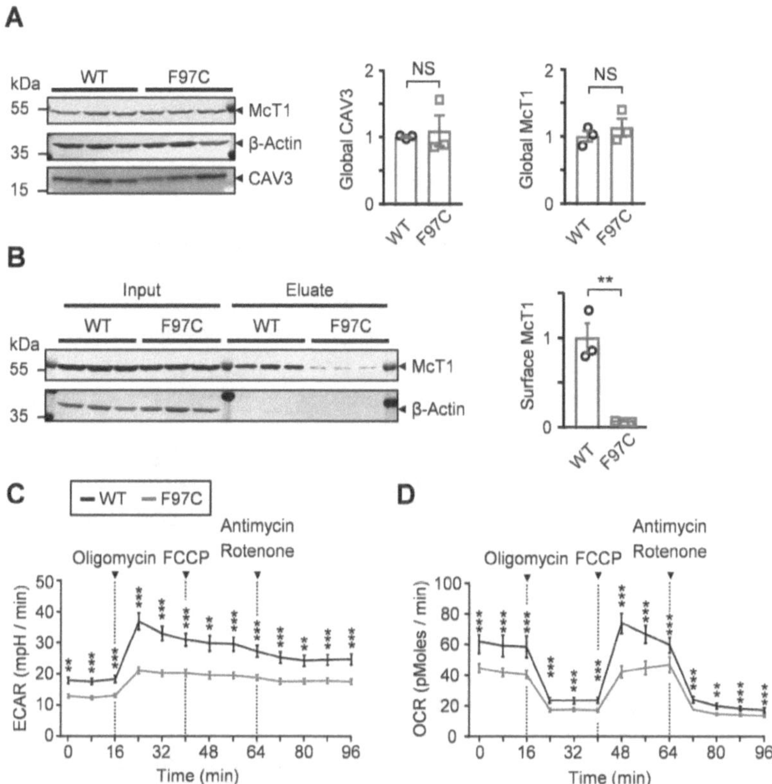



Figure 3.8 **F97C disrupts McT1 function in human cardiomyocytes. A**, Immunoblot analysis of human iPSC-derived F97C knockin cardiomyocytes. F97C CAV3 and McT1 were robustly expressed. Bar graph showing no significant differences between F97C and WT cardiomyocytes normalized to β-Actin. n=3. **B**, Extracellular protein biotinylation was used in living human cardiomyocytes to assess McT1 surface expression. Biotinylated proteins were enriched by pull-down and McT1 identified by immunoblotting in the eluated fraction. Vice versa, β-Actin immunoblotting was used as negative cytosolic protein labeling control. Bar graph showing a significant loss of surface McT1 in F97C versus WT cardiomyocytes. n=3. Student's t-test, ** p<0.01. **C**, The extracellular acidification rate (ECAR) was blunted by F97C at baseline, which was exacerbated by oligomycin treatment. n=32. Student's t-test, *** p<0.001. **D**, The oxygen consumption rate (OCR) of human cardiomyocytes was significantly decreased by F97C at baseline. Significant differences were maintained each after treatment by oligomycin, FCCP, and Antimycin+Rotenone. n=38. Student's t-test, *** p<0.001.

### 3.11 Supplemental methods

### Detailed methods

### Supplemental figures and figure legends



### Supplemental tables and supporting information



### Supplemental references

## Detailed methods

**Human stem cell study ethical approval.** The study was approved by the Ethics Committee of University Medical Center Göttingen (approval number 10/9/15) and carried out in accordance with the approved guidelines. Written informed consent was obtained from all donors prior to participation in the study.

**CAV1 knock-out mouse study approval.** Breeding and humane euthanasia for organ harvesting were carried out according to guidelines for the care and use of laboratory animals, following directive 2010/63/EU of the European Parliament and in keeping with NIH guidelines. All procedures were reviewed by the institutional animal committee of the University Medical Center Göttingen and approved by the veterinarian state authority (LAVES, Oldenburg, Germany; 33.9-42502-04-18/2975).

**APEX2 plasmids and recombinant adenoviral vectors for cardiomyocyte transfection.** The V5 epitope and APEX2 were N-terminally tagged to wild-type mouse Caveolin3 (V5-APEX2-CAV3). In addition, plasmids for expression of V5-APEX2-CAV3 with the CAV3 missense mutations F97C or S141R were generated. The CAV3 cDNA (MR226246, Origene) was amplified according to the manufacturer's instructions (In-Fusion HD Cloning Kit, Clontech) using the following primers:
- Fwd-5'-GGGGCAGCGGCTCGAGCATGATGACCGAAGAGCACA-3'
- Rev-5'-TAGATGCATGCTCGAGTTAGCCTTCCCTTC-3'

The CAV3 insert was cloned into the V5-APEX2 tagged pcDNA3 vector using XhoI restriction digest (NEB). The construct was transformed in Stellar Competent Cells (Clontech). The CAV3 mutations F97C or S141R were introduced by site-directed mutagenesis (Gene Art Site-Directed Mutagenesis PLUS Kit, Thermo Fisher Scientific) using the following primers:
- Fwd-5'-TCGCCTGTATCTCCTCTTGCCACATCTGGGC-3'
- Rev-5'-GCCCAGATGTGGCAAGAGGAGATACAGGCGA-3'

The resulting construct was transformed in DH5α cells (One Shot MAX Efficiency DH5α-T1R, Thermo Fisher Scientific). For adenoviral transfection, a custom-designed bicistronic subtype-5 vector (pO6A5-CMV, Sirion Biotech) was used to express eGFP and V5-APEX2 or V5-APEX2-CAV3 in neonatal cardiomyocytes. In analogy, adenoviral vectors of V5-APEX2-CAV3 containing the CAV3 mutations F97C or S141R were prepared and transfected. Adenoviral transduction of cardiomyocytes was monitored by eGFP fluorescence (Axiovert A1, Zeiss).

**Neonatal rat cardiomyocyte (NRCM) isolation and purification.** Hearts from 40 Wistar rat pups (P0-P3) were collected on ice in CBFHH buffer (Supplement Table 3.2). The atria were excised with scissors (914012-12, FST) under magnification view, the ventricles harvested for digestion (Enzyme D, Neonatal Heart Dissociation Kit, Miltenyi Biotech) and dissociation (gentleMACS Dissociator, Miltenyi Biotech)  according to the manufacturer's instructions (Neonatal Heart Dissociation Kit, Miltenyi Biotech). To enrich isolated ventricular cardiomyocytes, the raw cell suspension was filtered by gravity through a stainless steel mesh (grid size 250 µm, Thermo Fisher Scientific). The cells were pelleted by centrifugation (60 x g for 20 min at 4 °C), resuspended in 5 mL ice-cold PBS (PBS, pH 7.4, without $Ca^{2+}$and $Mg^{2+}$, Gibco), and each 2.5 mL of the suspension was layered on top of two Percoll density gradients (63 %, 40.5 % Centrifugation Media, pH 8.5 to 9.5, GE-Healthcare) using published protocols,[1] and centrifuged at 3,000 x g for 30 min at RT (acceleration speed 9; deceleration speed 0; Heraeus Multifuge X1R, Thermo Fisher Scientific). NRCM enriched at the Percoll layer interface were collected with a 10 mL glass pipette (10 mL wide tip, Ratiolab) and suspended in 10 mL of 37 C NRCM cultivation medium (Supplement Table 3.2).

**3-state SILAC cardiomyocyte culture conditions.** SILAC containing DMEM (Flex Media, Gibco) without L-arginine and L-lysine was supplemented with penicillin-streptomycin (100 U/mL), D-glucose (1 g/L), Na-pyruvate (100 mM), and BRDU (10 mM) containing either heavy, medium, or light isotope lysine and arginine as follows: for *heavy* SILAC labeling, L-lysine [$^{13}C_6$, $^{15}N_2$]HCl (Lys-8) and L-arginine [$^{13}C_6$, $^{15}N_4$]HCl (Arg-10) were added; for *medium* SILAC labeling, L-lysine-4,4,5,5-$d_4$ (Lys-4) and L-arginine [$^{13}C_6$]HCl (Arg-6); and for *light* SILAC labeling, DMEM liquid medium with 1 g/L D-glucose was used. All solutions were vacuum-filtered (Steritop, Merck). For NRCM culture, 10 % (vol/vol) heat inactivated FBS (Gibco) was added to the medium. NRCMs were seeded at a density of 500,000 cells on 35 mm dishes (CELLSTAR 6-well plate, Greiner) coated with collagen (13.96 mg/mL Collagen I rat tail, Corning) diluted 1:100 in PBS (PBS, pH 7.4, without $Ca^{2+}$ and $Mg^{2+}$, Gibco) and cultivated for 13 days in 2 mL light (non-labeled), medium, or heavy SILAC medium in 5 % $CO_2$ / 21 % $O_2$ at 37 °C (Heracell VIOS, Thermo Fisher Scientific). SILAC media were completely exchanged every 2nd day. Mass spectrometry determined SILAC incorporation (%) for up to 20 days in culture. SILAC incorporation reached a plateau (>95%) after 13 days culture (Supplement Figure 3.10 A).
For the proximity proteomic analysis (described in the next chapter) we used an experimental design for systematic label switching with three biological replicates as outlined in Figure 3.2 A.

**Ratiometric APEX2 mediated biotinylation in NRCM.** For ratiometric APEX2 mediated biotinylation of endogenous NRCM proteins, SILAC labeled NRCM were transfected with recombinant adenoviral vectors expressing V5-APEX2-CAV3 for 48 h using MOI 1 between day 11 and 13 in SILAC culture (for higher MOI doses please see Supplement Figure 3.9 A). In parallel, adenoviral vectors expressing soluble V5-APEX2 or eGFP were used as controls. Based on protocols published previously for ratiometric APEX biotinylation in heterologous cell systems,[2,3] 1 mL of each SILAC medium (chapter above) was exchanged by the same SILAC medium containing 500 µM biotin-phenol. After 30 min equilibration, a final concentration of 1 mM $H_2O_2$ was added and the medium gently mixed for 1 minute. After 1 min, the biotinylation reaction was quenched by replacing the medium with 1 mL quenching buffer (Supplement Table 3.3). NRCM were washed thrice with quenching buffer, scraped (Cell Scraper 25 cm, Sarstedt), and collected in 250 µL RIPA buffer (Supplement Table 3.3). The NRCM suspension was passed 15 times through a 27 gauge syringe on ice and centrifuged at 13,000 x g for 10 min at 4 °C to collect the solubilized proteins in the supernatant. The protein concentrations were determined by absorption measurement (Pierce 660 nm protein assay, Thermo Fisher Scientific). Heavy, medium and light labeled NRCM lysates were mixed at a 1:1:1 ratio at a total protein concentration of 250 µg.

For ratiometric APEX2 mediated biotinylation for the CAV3 mutations F97C and S141R, NRCM were transfected with adenoviral vectors containing mutant V5-APEX2-CAV3 for 48 h at MOI 1.

**Avidin capture and elution of biotinylated proteins.** Avidin beads (Pierce Monomeric Avidin Agarose, Thermo-Fisher-Scientific) were equilibrated in a ratio 1:1 with RIPA buffer (Supplement Table 3.3) and 80 µL avidin beads added to 250 µg of NRCM lysate. The suspension was gently rotated for 1 h at 4 °C in a spin column (Pierce Spin Columns Screw Cap, Thermo Fisher Scientific). Next, beads were washed twice with 500 µL RIPA buffer, once with 500 µL Tris/HCl buffer containing 2 M urea (pH 8.0) and again twice with 500 µL RIPA quenching buffer (Supplement Table 3.3). Two centrifugation steps at 100 x g for 30 sec and 2 min at 2000 x g (Heraeus, Fresco 21 centrifuge, Thermo Fisher Scientific) were used to harvest the beads, while the supernatant was discarded. Biotinylated proteins were eluted in 75 µL biotin buffer (Supplement Table 3.3) for 15 min at RT, followed by 15 min at 70 °C. Beads were pelleted by centrifugation for 1 min at 1000 x g and the supernatant containing the eluted proteins was collected. The eluted proteins were analyzed by mass spectrometry as described below.

**Sample preparation for NanoLC-MS/MS analysis of SILAC labeled samples.** Mass spectrometry was performed by the proteomic service unit in Göttingen according to published protocols.[4] Eluted protein samples were fractionated on 4-12 % Bis-Tris minigels (NuPAGE Novex, Invitrogen). Gels were stained with Coomassie Blue overnight (Coomassie Brilliant Blue R-250 Staining Solution, BioRad) for protein visualization, and each lane sliced into 11 equal-sized gel pieces. After washing the gel pieces with 50 mM ammonium bicarbonate (TEAB, Sigma-Aldrich), gel slices were reduced with 10 mM dithiothreitol (1,4-dithiothreitol, Sigma-Aldrich), alkylated with 55 mM iodoacetamide (2-iodoacetamide, Sigma-Aldrich), and digested with endopeptidase trypsin (sequencing grade, Promega) diluted 1:50 in 55 mM iodoacetamide overnight. Post-trypsin peptides were solubilized in MS loading buffer (Supplement Table 3.4), dried (SpeedVac, Thermo Fisher Scientific), reconstituted in MS loading buffer and prepared for NanoLC-MS/MS analysis as described previously.[5]

**NanoLC-MS/MS analysis of SILAC-labeled samples.** NanoLC-MS/MS analysis was performed by the proteomic service unit in Göttingen according to published protocols.[4] For mass spectrometric analysis of solubilized trypsin peptides, samples were enriched on a self-packed reversed phase-C18 precolumn (0.15 mm ID x 20 mm, Reprosil-Pur120 C18-AQ 5 µm, Dr. Maisch, Ammerbuch-Entringen, Germany) and separated on an analytical reversed phase-C18 column (0.075 mm ID x 200 mm, Reprosil-Pur 120 C18-AQ, 3 µm, Dr. Maisch) using a 30 min linear gradient of 5-35 % acetonitrile/0.1 % formic acid (v/v) at 300 nl min$^{-1}$. The eluent was analyzed on a mass spectrometer (Q Exactive hybrid quadrupole/orbitrap, Thermo Fisher Scientific) equipped with a FlexIon nanoSpray source and operated under Excalibur 2.5 software using a data-dependent acquisition method. Each experimental cycle was of the following form: one full MS scan across the 350-1600 $m/z$ range was acquired at a resolution setting of 70,000 FWHM, and AGC target of 1*10e6 and a maximum fill time of 60 ms. Up to the 12 most abundant peptide precursors of charge states 2 to 5 above a 2*10e4 intensity threshold were then sequentially isolated at 2.0 FWHM isolation width, fragmented with nitrogen at a normalized collision energy setting of 25%, and the resulting product ion spectra recorded at a resolution setting of 17,500 FWHM, and AGC target of 2*10e5 and a maximum fill time of 60 ms. Selected precursor $m/z$ values were then excluded for the following 15 s. Two technical replicates per sample were acquired.

**APEX2 assay data processing.** Raw data were processed using quantitative proteomic software (MaxQuant Software version 1.5.7.4, Max Planck Institute for Biochemistry). Proteins were identified against a UniProtKB-derived *rattus norvegicus* protein sequence database (v2018.02, 37830 protein entries) along with a set of common lab contaminants. The search was performed with trypsine as enzyme and iodoacetamide as cysteine blocking agent. Up to two missed tryptic cleavages and methionine oxidation as a variable modification were allowed for. Instrument type 'Orbitrap' was selected to adjust for MS acquisition specifics. The Arginine Arg-10, Arg-6 and Lysine Lys-8, Lys-6 labels including the 'Re-quantify' option were specified for relative protein quantitation. For identification of APEX2 biotinylated proteins (each for the WT, F97C, or S141R forms of V5-APEX2-CAV3), the ratios of V5-APEX2-CAV3 versus V5-APEX2 or eGFP were calculated and log2 transformed. The V5-APEX2-CAV3 / V5-APEX2 ratio was plotted on the X-axis and the V5-APEX2-CAV3 / eGFP ratio on the Y-axis (Figure 3.2 D, Figure 3.7 A and 3.7 C). Scatter plots were generated with Prism version 7.03 (GraphPad). Enriched biotinylated proteins were tested for statistical significance (p<0.05) by one sample z-test (Excel2007, Microsoft Office) and visualized as 'positive' or 'negative' hits including proteins-of-interest (POIs). See Excel file for mass spectrometry results (Table 8.11, 8.12 and 8.13 (**see Appendix**)).

**Immunoblotting and streptavidin blotting for protein analysis.** Mouse heart tissue, NRCMs, iPSC-cardiomyocytes, or HEK293A cells were homogenized in ice-cold RIPA buffer (Supplement Table 3.7) by 20 strokes on ice using a Potter homogenizer (RW20 digital, IKA). The homogenate was centrifuged at 10,000 x g for 10 min at 4 °C to pellet insoluble materials and protein concentration determined (Pierce BCA Protein Assay Kit; Thermo Fisher Scientific). For immunoblotting, 30 µg of cleared homogenate was loaded per lane onto a 4-20 % Tris-Glycine gradient gel (Novex 4-20 % Tris-Glycine, Thermo Fisher Scientific) and resolved by SDS gel electrophoresis at constant 200 V for 45 min. Proteins were transferred onto PVDF membranes (0.45 mm, Immobilon-FL, Merck Millipore) using an electrophoretic transfer cell (Mini Trans-Blot Electrophoretic Transfer Cell, Bio-Rad) at constant 100 V for 1 h in transfer buffer (Supplement Table 3.7) at 4 °C. PVDF membranes were blocked for 1 h in 5 % w/v non-fat milk (Milkpowder, Roth) in Tris-buffered saline with 0.05 % v/v Tween (Tween 20, Sigma Aldrich). PVDF membranes were incubated with the primary antibodies (Table 8.10 (**see Appendix**)) at 4 °C overnight, washed thrice with PBS (pH 7.4, without $Ca^{2+}$ and $Mg^{2+}$, Gibco) and incubated with fluorescent anti-mouse or anti-rabbit secondary antibodies at a dilution of 1:15,000 for a minimum period of 1 h at RT (P/N 926-32212, P/N 926-68072, P/N 926-32213, P/N 926-68073, IRDye LI-COR). Fluorescence signals were captured with the Odyssey CLx imaging system (LI-COR) and band intensities analyzed with Image Studio Lite Version 5.2 (LI-COR).

To analyze biotin-phenol labeled proteins (Figure 3.2 C and Supplement Figure 3.9 B-C) PVDF membranes were incubated with streptavidin (RD680, LI-

COR) for at least 1h at RT and the fluorescence detected with the Odyssey CLx imaging system (LI-COR) as described above.

**Blue Native (BN)-PAGE analysis.** BN-PAGE was used to analyze oligomeric complexes of endogenous CAV3 with V5-APEX2-CAV3 transfected in NRCM (Figure 3.1 D) as well as V5-APEX2-CAV3, V5-APEX2-CAV3-F97C or V5-APEX2-CAV3-S141R each transfected in HEK293A cells (Figure 3.6 B). Transfected cells were centrifuged at 13,000 x g for 10 min at 4 °C and 100 mg of the cell pellet resuspended in 1 mL homogenization buffer (Supplement Table 3.10). Cells were homogenized at 4 °C by 50 strokes on ice using a Potter homogenizer (RW20 digital, IKA). Homogenates were centrifuged at 1,000 x g for 10 min at 4 °C to remove cell debris. The cleared supernatant was centrifuged at 100,000 x g for 1 h (Optima Max-XP, MLA-150 rotor, Beckman) to enrich the membrane fraction. The plasma membrane fraction was resuspended in 30 µL solubilization buffer (Supplement Table 3.10), snap-frozen and stored at -80 °C. Solubilized membranes were thawn on ice and the protein concentration determined by absorption measurement (Pierce BCA Protein Assay Kit; Thermo Fisher Scientific). Digitonin (Digitonin, Sigma Aldrich) was added as detergent (6 g digitonin / g protein) and insoluble membranes were removed by centrifugation at 13,000 x g for 10 min at 4 °C. The cleared supernatant was mixed 1:10 with a Coomassie blue solution (Coomassie Brilliant Blue R-250, 5 % w/v, Sigma Aldrich) and a glycerol solution (Glycerol, 50 % w/v, Sigma Aldrich). Anode/cathode buffers were prepared according to manufacturer's instructions (NativePAGE Bis-Tris Mini Gel Electrophoresis Protocol, Thermo Fisher Scientific). For BN-Page, 50 µg of solubilized membrane proteins were separated on a 3-12% Bis-Tris gradient gel (NativePAGE 3-12% Bis-Tris Gel, Thermo Fisher Scientific) at constant 150 V for 1 h, followed by replacing the cathode buffer (Dark Blue Cathode Buffer, Novex) to cathode buffer light (Light Blue Cathode Buffer, Novex) and electrophoresis at constant 250 V for 1 h. Native markers (Serva Native Marker, Serva) were used to estimate molecular weight. Solubilized membrane proteins were transferred onto PVDF membranes (0.45 mm, Immobilon-FL, Merck Millipore) using an electrophoretic transfer cell (Mini Trans-Blot Electrophoretic Transfer Cell, Bio-Rad) at constant 50 V for 2 h in transfer buffer (Supplement Table 3.7) at 4 °C. PVDF membranes were blocked for 1 h in 5 % w/v non-fat milk (Milkpowder, Roth) in Tris-buffered saline with 0.05 % v/v Tween (Tween 20, Sigma Aldrich). Immunoblotting (Figure 3.1 D and Figure 3.6 B) was performed using V5 and CAV3 antibodies (Table 8.10 (**see Appendix**)).

**Co-immunoprecipitation of CAV interacting proteins.** Mouse ventricular heart lysates were solubilized with CHAPS co-IP buffer (Supplement Table 3.13) and 500 µg was incubated with 3 µg anti-CAV3 antibody (ab2912,

Abcam), 3 µg anti-CAV1 antibody (ab2910, Abcam) or normal rabbit IgG (12-370, Merck) antibody at 4 °C overnight. The samples were incubated with magnetic beads (Dynabeads Protein G, 15 µL, Thermo Fisher Scientific) in 100 µL CHAPS co-IP buffer for 2 h at 4 °C. The magnetic beads were extracted with a magnet (DynaMag-2 Magnet, Thermo Fisher Scientific), the solution discarded and the beads washed thrice with 500 µL ice-cold CHAPS co-IP buffer to minimize unspecific binding. Precipitated proteins were eluted in 60 µL of 2 × SDS buffer containing β-mercaptoethanol (Supplement Table 3.7). For the McT1 co-IP, 60 µL of 2x LDS buffer without β-mercaptoethanol (1× NuPAGE, Invitrogen) was used to decrease IgG signals at 55 kDa. Eluated samples were heated to 70 °C for 5 min and resolved on 4-20% Tris-Glycine gradient gels (Novex 4-20% Tris-Glycine, Thermo Fisher Scientific). For SDS gel electrophoresis and protein transfer please see: **Immunoblotting and streptavidin blotting for protein analysis**. After transfer and blocking, primary antibodies against the proteins shown in Figure 3.3 E and Figure 3.4 C were applied (Table 8.10 (**see Appendix**)) at 4 °C overnight. To reduce unspecific background signals, for Aquaporin1, McT1, Ncx1, and TfR1 antibody incubation the IRDye-680 detection reagent was added according to the manufacturer's instructions (Quick Western Kit, LI-COR).

**Co-immunoprecipitation of V5-APEX2-CAV3.** Adenovirally transfected NRCMs expressing V5-APEX2-CAV3, V5-APEX2-CAV3-F97C, V5-APEX2-CAV3-S141R or V5-APEX2 were solubilized with sodium deoxycholate co-IP buffer (Supplement Table 3.13) and incubated with 3 µg anti-V5 antibody (R960-25, Thermo Fisher Scientific) at 4 °C overnight. Magnetic beads (Dynabeads Protein G, Thermo Fisher Scientific) were added to the sample, incubated for 2 h at 4 °C, washed and resuspended in 2× SDS buffer containing β-mercaptoethanol as described above. Samples were heated to 70 °C for 5 min and resolved on 4-20% Tris-Glycine gradient gels (Novex 4-20% Tris-Glycine, Thermo Fisher Scientific). For SDS gel electrophoresis and protein transfer please see: **Immunoblotting and streptavidin blotting for protein analysis**. After transfer on PVDF membranes and blocking, PVDF was incubated with the antibodies anti-V5 (R960-25, Thermo Fisher Scientific) and anti-CAV3 (ab2912, Abcam) (Figure 3.1 C).

**Sample preparation for label-free SWATH-MS (Sequential Window Acquisition of All THeoretical Mass Spectra).** Label-free SWATH-MS quantification was performed according to published protocols.[5] Samples were run on 4-12% NuPAGE Novex Bis-Tris Minigels (4-12% NuPAGE, Invitrogen) for a short distance (~1 cm), cut out as a single fraction and trypsinized as described: **In-gel tryptic digestion**. Post-trypsin peptides were solubilized in MS loading buffer (Supplement Table 3.4), dried (SpeedVac, Thermo Fisher

Scientific), reconstituted in MS loading buffer, and prepared for nanoLC-MS/MS as described previously.[5] A synthetic peptide standard for retention time alignment was used to spike all samples (iRT Standard, Biognosys).

Affinity purification (AP) followed by label-free quantification (AP-MS) was performed as previously described with few modifications.[6] CAV1 and CAV3 were immunoprecipitated from 500 µg mouse ventricular tissue. Normal rabbit IgG (12-370, Merck) was used as negative control. Immunoprecipitates were run on a 4-12% NuPAGE Novex Bis-Tris Minigels (4-12% NuPAGE, Invitrogen) as a single fraction and prepared for nanoLC-MS/MS as described above.

**NanoLC-MS/MS analysis by label-free SWATH-MS.** Label-free SWATH-MS quantification was performed by the proteomic service unit in Göttingen according to published protocols.[5] Protein digests were analyzed on a nanoflow chromatography system (Eksigent nanoLC425, SCIEX) hyphenated to a hybrid triple quadrupole-TOF mass spectrometer (TripleTOF 5600+, SCIEX) equipped with a Nanospray III ion source (Ionspray Voltage 2400 V, Interface Heater Temperature 150 °C, Sheath Gas Setting 12) and controlled (Analyst TF 1.7.1 software build 1163, SCIEX). Peptides were dissolved in MS loading buffer (Supplement Table 3.4) to a concentration of 0.3 µg/µL. For each analysis 1.5 µg of digested protein were enriched on a precolumn (0.18 mm ID x 20 mm, Symmetry C18, 5 µm, Waters) and separated on an analytical RP-C18 column (0.075 mm ID x 250 mm, HSS T3, 1.8 µm, Waters) using a 90 min linear gradient of 5-35% acetonitrile/0.1% formic acid (v/v) at 300 nl min$^{-1}$.

Qualitative LC/MS/MS analysis was performed using a Top25 data-dependent acquisition method with an MS survey scan of $m/z$ 350–1250 accumulated for 350 ms at a resolution of 30,000 full width at half maximum (FWHM). MS/MS scans of $m/z$ 180–1600 were accumulated for 100 ms at a resolution of 17,500 FWHM and a precursor isolation width of 0.7 FWHM, resulting in a total cycle time of 2.9 s. Precursors above a threshold MS intensity of 125 cps with charge states 2+, 3+, and 4+ were selected for MS/MS, the dynamic exclusion time was set to 30 s. MS/MS activation was achieved by collision-induced dissociation using nitrogen as a collision gas and the manufacturer's default rolling collision energy settings. Two technical replicates per sample were analyzed to construct a spectral library.

For quantitative SWATH analysis, MS/MS data were acquired using 65 variable size windows[7] across the 400-1,050 $m/z$ range. Fragments were produced using rolling collision energy settings for charge state 2+, and fragments acquired over an $m/z$ range of 350–1400 for 40 ms per segment. Including a 100 ms survey scan this resulted in an overall cycle time of 2.75 s. 3x3 replicates (biological x technical) were acquired for each biological state.

**Data processing for label-free SWATH-MS.** Data processing was performed by the proteomic service unit in Göttingen according to published protocols.[8] Protein identification was achieved (ProteinPilot Software version 5.0 build 4769, SCIEX) at "thorough" settings. The combined qualitative analyses were searched against the UniProtKB mouse reference proteome (revision 04-2018, 61,290 entries) augmented with a set of 52 known common laboratory contaminants to identify proteins at a False Discovery Rate (FDR) of 1%. Spectral library generation and SWATH peak extraction were achieved (PeakView Software version 2.1 build 11041, SCIEX) using the SWATH quantitation microApp version 2.0 build 2003. Following retention time correction using the iRT standard, peak areas were extracted using information from the MS/MS library at an FDR of 1%.[6] The resulting peak areas were then summed to peptide and finally protein area values per injection, which were used for further statistical analysis.

At least, three biological replicates were performed and proteomic differences were evaluated for statistical significance (p<0.05) by permutation-based false-discovery rate analysis (t-test, p>0.05, FDR=5%, S0=0.1). Furthermore, a $\log_2$ fold change ratio ≥1 was used as cutoff.[9] CAV1 versus CAV3 interacting proteins (Figure 3.4 A) identified by SWATH-MS were illustrated by volcano plot (Perseus, MaxQuant). The fold changes were log2 transformed and plotted on the X-axis, while the permutation-based false-discovery rate analysis p values (t-test, p>0.05, FDR=5%, S0=0.1, Perseus, MaxQuant) were $-\log_{10}$ transformed and plotted on the Y-axis. Putative binding partners are listed in Table 8.14 and 8.15 (**see Appendix**).

**STRING analysis.** Following protein identification by MS, we analyzed their context based on the STRING database for cellular compartments (Supplement Figure 3.10 B) and protein–protein interaction networks (string-db.org) using Gene Ontology (GO) terms (Figure 3.2 E and Supplement Figure 3.15 D-E) as described previously.[10] We used a scoring cut-off of ≥0.75 to define positive interactions following published workflows.[11]

**Cell culture of human induced pluripotent stem cells (iPSCs).** Cell culture and ventricular differentiation of human induced pluripotent stem cells (iPSCs) was performed by the Stem Cell Unit in Göttingen. The human iPSC lines isWT1.14 (UMGi014-C.14; abbreviated as WT iPSC), isWT1-CAV3-KO.34 (UMGi014-C-3.34; abbreviated as CAV3 KO iPSC) and isWT1-CAV3-F97C.56 (UMGi014-C-4.56; abbreviated as F97C KI iPSC) were maintained on Matrigel-coated (Matrigel, growth factor reduced, BD Biosciences) 35 mm plates (CELLSTAR 6-well plate, Greiner), passaged every 4-6 days with a non-enzymatic cell dissociation reagent (Versene solution, Thermo Fisher Scientific) and cultured in iPSC medium (Supplement Table 3.5) for 24 h after passaging and with daily complete medium changes. For cell culture a humidified incubator with 5% $CO_2$ and 21% $O_2$ at 37 °C was used (Heracell VIOS, Thermo Fisher Scientific).

**Stem cell differentiation.** Directed differentiation of human iPSCs into ventricular iPSC cardiomyocytes was performed via WNT signaling modulation as described previously.[12] The ventricular differentiation was initiated at 80%–90% confluence on Matrigel-coated 35 mm plates using the cardiac differentiation medium (Supplement Table 3.5) and sequential treatment with 4 µM of a GSK-3α/β inhibitor (CHIR-99021, Merck Millipore) for 48 hours, followed by 5 µM PORCN Inhibitor (IWP2, Merck Millipore) for 48 hours. The complete medium was replaced by cardio culture medium (Supplement Table 3.5) at day 8. Differentiated cultures around day 15 were digested with 0.25 % trypsin (Trypsin/EDTA, Thermo Fisher Scientific) and replated in 35 mm plates (CELLSTAR 6 Well plate, Greiner). Metabolic ventricular cardiomyocyte selection was done with cardio selection medium (Supplement Table 3.5) for 5 days. Afterwards, iPSC cardiomyocytes were cultured in cardio culture medium (Supplement Table 3.5) at least to day 60 for further maturation. iPSC cardiomyocytes were washed thrice with PBS (PBS, pH 7.4, without $Ca^{2+}$ and $Mg^{2+}$, Gibco) and scraped (Cell Scraper 25 cm, Sarstedt) in 500 µL ice cold PBS. iPSC cardiomyocytes were pelleted at 13,000 x g for 10 min at 4 °C, snap-frozen in liquid N2 and stored at -80 °C until further use.

**CRISPR/Cas9-mediated genome editing.** For genome editing of human iPSC lines we used ribonucleoprotein (RNP)-based CRISPR/Cas9 to target exon 1 of the human CAV3 gene to generate CAV3 knock-out iPSCs or exon 2 to introduce the F97C-CAV3 mutation (Supplement Figure 3.13 A and B). The guideRNA target sequences, with PAM in bold, were:
- CRISPR#1 5'-TCCCCCCAGCTCTGCGATGA**TGG**-3'
- CRISPR#2 5'-CACCGCCCAGATGTGGCAGA**AGG**-3'

For F97C-CAV3 knock-in via homology-directed repair, a single-stranded oligonucleotide with 60 bp homology arms including the respective SNP and silent SNPs for PAM mutation was used. The human iPSC line is WT1.14 (UMGi014-C.14; abbreviated as WT iPSC) was cultured in StemFlex medium (StemFlex medium, Thermo Fisher Scientific) on Matrigel-coated (growth factor reduced, BD Biosciences) plates and transfected by nucleofection according to the manufacturer's instructions (P3 Primary Cell 4D-Nucleofector X Kit, Lonza) between passage 12 to 15. The CRISPR/Cas9 RNP complex was assembled by mixing of the individual Alt-R CRISPR-Cas9 crRNA and the Alt-R CRISPR-Cas9 tracrRNA, preassembled in a 1:1 ratio, with the Alt-R S.p. HiFi Cas9 Nuclease 3NLS (all: IDT DNA Technologies) at a 1:3 molar ratio, incubated for 10 min at RT and diluted in nucleofector solution (P3 4D-Nucleofector X

Solution, Lonza). 1 h before nucleofection, iPSCs were pretreated with 2 µM of Rho inhibitor (Thiazovivin, Merck Millipore) and dissociated using a non-enzymatic cell dissociation reagent (Versene solution, Thermo Fisher Scientific) at a confluence of 70-80%. For each approach, $2\times10^6$ iPSCs, quantified by cell counter (CASY, OMNI Life Science), were used according to the manufacturer's instructions. Following nucleofection, iPSCs were replated in two Matrigel-coated (Matrigel, growth factor reduced, BD Biosciences) wells of a 35 mm plates (CELLSTAR 6 Well plate, Greiner) and cultured in StemFlex medium (Thermo Fisher Scientific) supplemented with 2 µM of Rho inhibitor (Thiazovivin, Merck Millipore). After 48 h, transfected iPSCs were replated as single cells by limited dilution as described previously,[12] and cultured in StemFlex medium (Thermo Fisher Scientific) for one week. Individual iPSC colonies were manually picked and expanded for approximately one to three weeks in StemMACS iPS-Brew XF medium (Miltenyi Biotech) with daily medium change. Expanded colonies were analyzed for genetic modification by Sanger sequencing (Supplement Figure 3.13 B and D) and positive clones were selected for further analysis. The normal karyotype was determined in post-edited cells and clones were resequenced for purity every 5-10 passages. Genomic stability of human iPSC cultures was assessed between passage 25 and 30 using the G-banding method according to previous protocols.[13] At least 15 metaphase cells per sample were analyzed, all of which were concluded to have no structural abnormality (Supplement Figure 3.13 E).

**Immunocytochemical staining and flow cytometry of stem cells.** Immunostaining and flow cytometry was performed by the Stem Cell Unit in Göttingen according to previous published protocols.[14] For immunocytochemical studies, cells were cultured on glass coverslips (Ø 18mm, width 1.5, Menzel), fixed (Roti-Histofix 4%, Carl Roth) at RT for 20 min, and blocked with stem cell blocking solution (Supplement Table 3.6) at 4 °C overnight. Cells were incubated with primary antibodies against OCT4, NANOG, and TRA-1-60 (Table 8.10 (**see Appendix**)) diluted in stem cell blocking solution (Supplement Table 3.6) at 4 °C overnight, washed thrice with stem cell blocking solution, and finally incubated with secondary antibodies in stem cell blocking solution at RT for 1 h. Cells were permeabilized with 0.1 % Triton-X100 (Triton-X100, Carl Roth) in 1 % BSA (BSA; Sigma-Aldrich) in PBS (PBS, pH 7.4, without $Ca^{2+}$and $Mg^{2+}$, Gibco). Nuclei were stained with 4.8 µM DAPI (DAPI solution, Thermo Fisher Scientific) at RT for 10 min. Samples were mounted (Fluoromount-G, Thermo Fisher Scientific) and images collected by light microscopy (Axio Imager M2 microscopy system, Zen 2.3 software, Carl Zeiss) (Supplement Figure 3.13 F).

For flow cytometry, cells were manually agitated with a 10 mL glass pipette (Serological Pipette 10 mL, Sarstedt), fixed (Roti-Histofix 4 %, Carl Roth) at RT for 20 min, and blocked with stem cell blocking solution (Supplement Table 3.6) at 4 °C for at least 2 h. iPSCs were permeabilized with 0.1% Triton-X100 (Triton-X100, Carl Roth) in stem cell blocking solution and co-incubated with fluorescence-conjugated antibodies against OCT4 and TRA-1-60 at RT for 1 h (Table 8.10 (**see Appendix**)). Nuclei were co-stained with 8.1 µM Hoechst (Hoechst 33342, Thermo Fisher Scientific). Subsequently, cells were analyzed using the flow cytometry (LSRII, BD Biosciences) using BD FACSDiva software (BD Biosciences). Gating of cells was applied based on forward scatter area (FSC-A) and sideward scatter area (SSC-A) as well as on gating of single cells based on DNA signal width. At least 10,000 events were analyzed per sample.

**iPSC-cardiomyocyte 3-bromopyruvate (3-BP) uptake and cell viability assay.** Differentiated iPSCs cardiomyocytes cultured for 30 days from the WT and CAV3 KO iPSC lines were digested with 0.25 % trypsin (Trypsin/EDTA, Thermo Fisher Scientific) and 1 million iPSC-cardiomyocytes quantified by cell counter (CASY, OMNI Life Science). iPSC-cardiomyocytes were seeded on Matrigel-coated (growth factor reduced, BD Biosciences) 35 mm dishes (CELLSTAR 6-well plate, Greiner). WT and CAV3 KO iPSC-cardiomyocytes were cultured in cardio culture medium (Supplement Table 3.5) until day 60 and the cardio culture medium exchanged by the same medium containing 50 µM 3-bromopyruvate (3-BP). After incubation for 3 h in 5 % $CO_2$ / 21 % $O_2$ at 37 °C (Heracell VIOS, Thermo Fisher Scientific) the culture medium was collected and extracellular release of the lactate dehydrogenase (LDH) into the media quantified by a coupled enzymatic reaction (LDH Cytotoxicity Assay Kit, Thermo Fisher Scientific). For this purpose, 50 µL of culture media were each transferred into a 96-well plate (Cellstar 96 well plates, Greiner) for triplicate measurements and the enzymatic reaction initiated by adding 50 µL reaction mixture (LDH Cytotoxicity Assay Kit, Thermo Fisher Scientific). After 30 min incubation, the absorbance was measured at 490 nm and 680 nm (Spark 10M, Tecan) and LDH release calculated according to the manufacturer's instructions (LDH Cytotoxicity Assay Kit, Thermo Fisher Scientific) (Figure 3.5 C).

**iPSC-cardiomyocyte cell surface biotinylation and elution of biotinylated surface proteins.** 60 day cultured human iPSC-cardiomyocytes were washed thrice with 500 µL PBS (PBS, pH 7.4, without $Ca^{2+}$ and $Mg^{2+}$, Gibco) to remove primary amine groups. For cell surface biotinylation, iPSC-cardiomyocytes were incubated for 1 h at 4 °C with 2 mM tagging solution (EZ-Link Sulfo-NHS-Biotin, Thermo Fisher Scientific) or PBS (PBS, pH 7.4, without $Ca^{2+}$ and $Mg^{2+}$, Gibco), the latter as negative control. Cell surface biotinylation was quenched after 1 h by adding 100 µL of 1 M Tris (pH 7.5) to the tagging solution and following incubating for 5 min at RT. The iPSC-cardiomyocytes were washed twice with ice cold PBS (PBS, pH 7.4, without $Ca^{2+}$and $Mg^{2+}$, Gibco), scraped (Cell

Scraper 25 cm, Sarstedt) in 250 µL ice cold PBS (PBS, pH 7.4, without $Ca^{2+}$ and $Mg^{2+}$, Gibco), and centrifuged at 13,000 x g for 1 min. The pellet was resuspended in 500 µL RIPA buffer (Supplement Table 3.7) and homogenized by 20 strokes on ice using a Potter homogenizer (RW20 digital, IKA). The homogenate was centrifuged at 10,000 x g for 10 min at 4 °C to remove insoluble contents and the protein concentrations determined by absorption measurement (Pierce BCA Protein Assay Kit; Thermo Fisher Scientific). Biotinylated surface proteins were precipitated by adding 40 µg avidin beads to 500 µg lysate for 1 h at 4 °C in a spin column (Pierce Spin Columns Screw Cap, Thermo Fisher Scientific). Next, beads were washed thrice with 500 µL RIPA buffer. Two centrifugation steps at 100 x g for 30 sec and 2 min at 2000 x g (Heraeus, Fresco 21 centrifuge, Thermo Fisher Scientific) were used to harvest the beads and the supernatant was discarded. Biotinylated proteins were eluted in 100 µL 2 x SDS buffer containing β-mercaptoethanol (Supplement Table 3.7). For Immunoblotting, 15 µg of input and 15 µL eluate sample were loaded onto a 4-20% Tris-Glycine gradient protein gel (Novex 4-20% Tris-Glycine, Thermo Fisher Scientific). For SDS gel electrophoresis and protein transfer please refer to **Immunoblotting and streptavidin blotting for protein analysis**. Immunoblotting (Figure 3.5 B and Figure 3.8 B) was performed with McT1 and β-Actin antibodies (Table 8.10 (**see Appendix**)).

**iPSC-cardiomyocyte Seahorse studies.** Initially, iPSC-cardiomyocytes were cultured for 7 days in cardio culture medium (Supplement Table 3.5) to form a confluent synchronously beating monolayer. After 60 days of cultivation, iPSC cardiomyocytes were prepared for metabolic studies by exchanging the medium for the Seahorse XF assay buffer (Supplement Table 3.8). To determine the respiratory capacity, 20,000 cells seeded per Matrigel-coated well (growth factor reduced, BD Biosciences) using a Seahorse 96-well plate (XF96 cell culture microplate, Agilent). The Oxygen Consumption Rate (OCR) and the Extracellular Acidification Rate (ECAR) were measured with a Seahorse Extracellular Flux Analyzer (XF96, Seahorse Bioscience). Periodic measurements of OCR and ECAR were repeated under basal conditions and after inhibition of the ATP synthase (Oligomycin, 3 µM, Sigma Aldrich), after mitochondrial oxidative phosphorylation uncoupling (Carbonyl cyanide-4-(trifluoromethoxy)phenylhydrazone (FCCP), 1 µM, Sigma Aldrich) and after ComplexI/ComplexIII inhibition (rotenone, 2 µM plus antimycin A, 1 µM, Sigma Aldrich) according to a previously published protocol[15] (Figure 3.5 D-E and Figure 3.8 C-D). Mitochondrial and glycolytic ATP levels were calculated based on experimentally determined OCR and ECAR values (Supplement Figure 3.16 A-B) according to the manufacturer's protocol (Quantifying Cellular ATP Production Rate Using Agilent Seahorse XF Technology, Agilent).

**Transfection and live-cell cross-linking of V5-APEX2-CAV3 or V5-APEX2-CAV3-F97C expressing HEK293A cells.** HEK293A cells were passaged using trypsin (Trypsin/EDTA solution, Sigma) and seeded at a density of 250,000 cells on 35 mm dishes (CELLSTAR 6-well plate, Greiner). HEK293A cells were cultured in HEK293 medium (Supplement Table 3.9) in 5% $CO_2$ / 21% $O_2$ at 37 °C (Heracell VIOS, Thermo Fisher Scientific). At 70% confluency, HEK293A cells were transfected by plasmid using Lipofectamin 3000 (Lipofectamin 3000 Transfection Reagent, Thermo Fisher Scientific). For this, the culture medium was exchanged using 1 mL of the following transfection reagents: 3 µg plasmid expression vector (V5-APEX2-CAV3 or V5-APEX2-CAV3-F97C) with 3 µL Lipofectamin 3000 reagent and 6 µL P3000 in DMEM with low glucose (Dulbecco's Modified Eagle's Medium low glucose, Sigma) without supplements. After 6 h incubation in 5% $CO_2$ / 21% $O_2$ at 37 °C (Heracell VIOS, Thermo Fisher Scientific), the transfection mix was exchanged against HEK293 culture medium. 24 h post-transfection, HEK293A cells were washed thrice with PBS (PBS, pH 7.4, without $Ca^{2+}$ and $Mg^{2+}$, Gibco) to treat primary amine groups for cross-linking.

For cross-linking, a 50 mM stock solution of disuccinimidyl suberate (DSS, Thermo Fisher Scientific) was prepared by dissolving 2 mg DSS in 108 µL dimethyl sulfoxide (DMSO, Thermo Fisher Scientific) and HEK293A cells were incubated at the final concentrations of 100 µM or 300 µM DSS or DMSO as negative control for 1 h at 4 °C. Cross-linking was quenched by adding 50 µL of 1 M Tris (pH 7.5) for 5 min at RT. The HEK293A cells were scraped (Cell Scraper 25 cm, Sarstedt) in 250 µL ice-cold PBS (PBS, pH 7.4, without $Ca^{2+}$ and $Mg^{2+}$, Gibco) and centrifuged at 13,000 x g for 1 min. The pellet was resuspended in RIPA buffer (Supplement Table 3.7) and homogenized by 20 strokes on ice using a Potter homogenizer (RW20 digital, IKA). The homogenate was centrifuged at 10,000 x g for 10 min at 4 °C to remove insoluble material and the protein concentration determined by absorption measurement (Pierce BCA Protein Assay Kit; Thermo Fisher Scientific). For immunoblotting, 30 µg protein were loaded per lane onto a 4-20% Tris-Glycine gradient protein gel (Novex 4-20% Tris-Glycine, Thermo Fisher Scientific,). For SDS gel electrophoresis and protein transfer please refer to **Immunoblotting and streptavidin blotting for protein analysis**. Immunoblotting (Figure 3.6 D) was performed with the documented CAV3 antibody materials (Table 8.10 (**see Appendix**)).

**CAV1 KO mouse model.** All animal procedures were performed according to institutional rules reviewed by IACUC of the University Medical Center Göttingen and approved by the veterinarian state authority (LAVES, Oldenburg, Germany; 33.9-42502-04-18/2975). CAV1 KO mice were purchased form Jackson Lab (B6.Cg-*Cav1*[tm1MLs]/J, 007083) and back-crossed into the

C57BL/6N background. We used adult mice of 12-14 weeks age and mixed genders.

Mice were anesthetized with 3 % isoflurane, euthanized by cervical dislocation, and the heart extracted following protocols for the humane use of laboratory animals based on approval by the institutional animal care and use committee (T2/11, Lehnart). For ventricular tissue preparation for biochemical analysis, mice were anesthetized and the hearts perfused as detailed in the next chapter for 2 min to clear blood cells. Ventricular tissue was manually dissected under a binocular microscope (Stemi 305, Zeiss), snap-frozen in liquid N2, cut into small pieces, and stored at -80 °C until further used for analysis.

**Adult mouse ventricular cardiomyocyte isolation.** We used our customized, published protocol for isolation of adult ventricular cardiomyocytes.[16] The proximal aorta was connected to a 21 gauge cannula and connected to a modified Langendorff perfusion setup.[17] Hearts were perfused by constant flow at 4 mL/min with a nominally $Ca^{2+}$ free perfusion buffer (Supplement Table 3.11) for 4 min at 37 °C, followed by collagenase containing buffer (Supplement Table 3.11) for another 9 min at 37 °C. The ventricles were dissected under a binocular microscope (Stemi 305, Zeiss) in 2 mL digestion buffer and digestion was stopped by adding 3 mL stopping buffer (Supplement Table 3.11). Isolated ventricular cardiomyocytes were washed twice with the stopping buffer, cells sedimented for 8 min by gravity at RT, the supernatant discarded and the cells resuspended. Cell quality was documented by transmitted light imaging (Zeiss LSM 710 and 880, Jena, Germany) using Fiji (https://imagej.net/Fiji) following criteria documented previously.[17]

**Confocal microscopy and superresolution STED immunofluorescence nanoscopy.** Isolated cardiomyocytes were plated on glass coverslips (Ø 18 mm, width 1.5 mm, Menzel) after coating with laminin (2 mg/mL) at a dilution of 1:10 in perfusion buffer (Supplement Table 3.11). Cardiomyocytes were fixed with 4% paraformaldehyde (PFA, Sigma-Aldrich) for 5 min at room temperature followed by three PBS (PBS, pH 7.4, without $Ca^{2+}$ and $Mg^{2+}$, Gibco) washing steps. Fixed samples were incubated overnight at 4 °C in blocking/permeabilization buffer followed by incubation with the primary antibodies (Table 8.10 (**see Appendix**)) diluted in blocking buffer overnight at 4 °C. After washing thrice with blocking buffer, samples were incubated with secondary antibodies diluted 1:1000 overnight at 4 °C. For confocal immunofluorescence microscopy, Alexa Fluor 633 and Alexa Fluor 514 conjugated antibodies were used (Table 8.10 (**see Appendix**)). For STED microscopy, STAR 635P and STAR 580 conjugated antibodies were used (Table 8.10 (**see Appendix**)). After washing thrice with PBS, samples were embedded in mounting medium with DAPI (ProLong Gold Antifade Mountant with DAPI, Thermo Fisher Scientific) for confocal microscopy or DAPI-free mounting medium (ProLong Gold Antifade Mountant, Thermo Fisher Scientific) for STED nanoscopy. Embedded samples were stored overnight at RT and imaged the next day.

Confocal images were acquired with a Zeiss LSM 710 microscope using a Plan-Apochromat 63x/1.40 oil objective and a pixel size of 50 x 50 nm. Alexa Fluor 514 was excited at 514 nm and detected at 520–620 nm. AlexaFluor 633 was excited at 633 nm and detected at 640–740 nm. The confocal laser power was adjusted to maximize resolution following established workflows.[17] For STED nanoscopy a Leica TCS SP8 system with a HC PL APO C2S 100x/1.40 oil objective and a pixel size 16.23 x 16.23 nm was used. STAR 635P was excited at 635 nm and detected at 650-700 nm. STAR 580 was excited at 580 nm and detected at 600-630 nm. For STED depletion of STAR 580 and STAR 635P a 775 nm laser beam was used. The STED laser power was adjusted to maximize resolution following previously established workflows.[17] Raw images were processed in Fiji (https://imagej.net/Fiji) following established protocols.[17]

**High-pressure freezing and electron tomography.** High-pressure freezing was performed at the EMBL Heidelberg electron microscopy core facility according to published protocols.[18] Isolated mouse ventricular cardiomyocytes were placed in 200 µm aluminium type A specimen carriers coupled with type B lids (HPF carrier, Leica) and the specimens were rapidly frozen (HPM100, Leica). Specimens were freeze substituted for 24 h in 1% OsO4 in acetone (AFS2, Leica), dehydrated in graded acetone and embedded (Epon-Araldite resin, EMS). Semi-thick (280 nm) sections were placed on formvar-coated slot-grids (TEM Grids, Science Services), post stained with 2% aqueous uranyl acetate (2% Uranyl Acetate Solution, Science Services) and Reynold's lead citrate (Supplement Table 3.12). Colloidal gold particles (Gold nanoparticles 15 nm, Sigma Aldrich) were added to both surfaces of the sections to serve as fiducial markers for tilt series alignment.

For imaging an intermediate voltage electron microscope (Tecnai TF30, FEI) was operated at 300 kV. Electron tomography was performed according to published protocols.[18] Images were captured on a 4K x 4K charge-coupled device camera (UltraScan, SerialEM software package; Gatan). For imaging, the specimen holder was tilted from +60 ° to -60 ° at 1 ° intervals. For dual-axis tilt series the specimen was then rotated by 90° in the X-Y plane, and another +60 ° to -60 ° tilt series was taken. The images from each tilt-series were aligned by fiducial marker tracking and back-projected to generate two single full-thickness reconstructed volumes (tomograms), which were then combined to generate a single high-resolution 3D reconstruction of the original partial cell volume.[19] Isotropic voxel size ranged from 0.765-1.206 nm. In some instances, tomograms were computed from montaged stacks, to increase the total reconstructed area to up to 10 µm x 10 µm in XY. Biologically meaningful resolution was approximately 4 nm in X-Y. All tomograms were processed and analyzed using IMOD software,[20] which was also used to generate 3D models of relevant structures of interest.[21] Models were smoothed and meshed to obtain the final 3D representation, in which spatial relations between caveolar and mitochondrial structures were quantified.

**Statistical analysis.** Data are presented as mean ± standard error of the mean (SEM) unless indicated otherwise. Unpaired 2-tailed Student's t-test or 1-way-ANOVA was applied as specified in the figure legends. A p value of less than 0.05 was considered statistically significant.

## Supplemental figures and figure legends

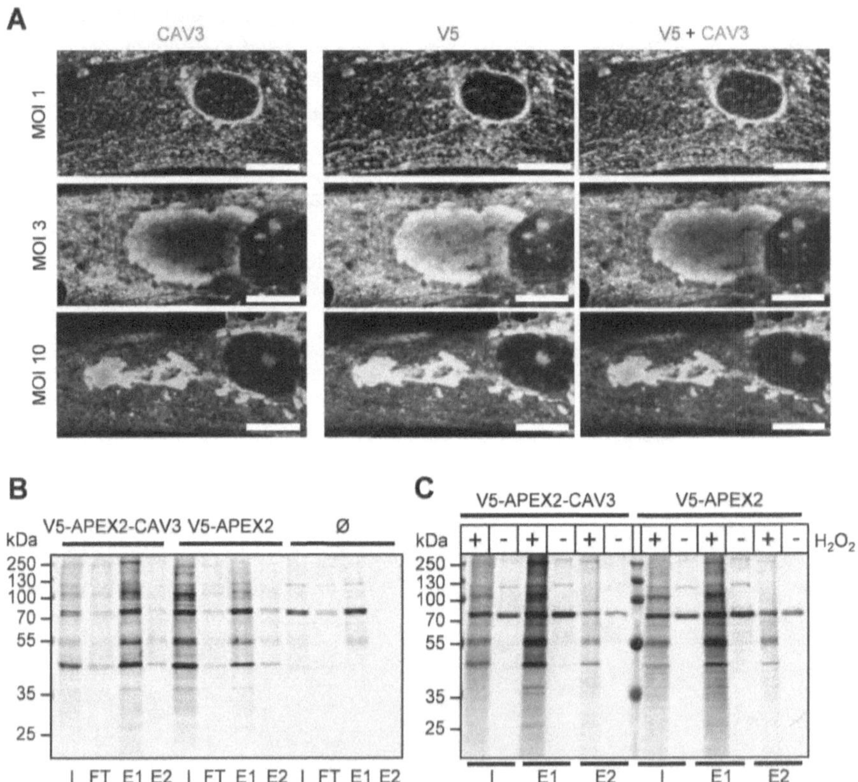


**Supplement Figure 3.9  CAV3 proximity labeling in living neonatal rat cardiomyocytes A**, Confocal microscopy showed the localization of V5-APEX2-CAV3 after adenoviral transfection at the indicated MOIs. Only a MOI of 1 resulted in subcellular V5-APEX2-CAV3 signals similar to endogenous CAV3. In contrast, MOIs of 3 or 10 resulted in abnormally large perinuclear, sharply demarked signal regions indicative of Golgi accumulation and a potential trafficking defect at MOI 10. Scale bars: 10 µm. **B**, APEX2 proximity labeling in V5-APEX2-CAV3 versus V5-APEX2 transfected or untransfected (Ø) NRCM. Biotinylated proteins were enriched by affinity purification and detected with streptavidin IRDye 680 RD. Dark signals indicate proteins eluted with biotin buffer (E1; Supplement Table 3.3), as second elution step (E2) with 2x SDS buffer confirmed sufficient elution in E1 (weaker signals in E2). Untransfected NRCM (Ø) were used as negative control and to document endogenously biotinylated proteins. I, input; FT, flow through; E, eluate; n=3. **C**, Biotinylation analyzed in V5-APEX2-CAV3 or V5-APEX2 transfected NRCM after 1-min treatment versus omission of $H_2O_2$ documenting APEX2-depedent biotinylated proteins. n=2.

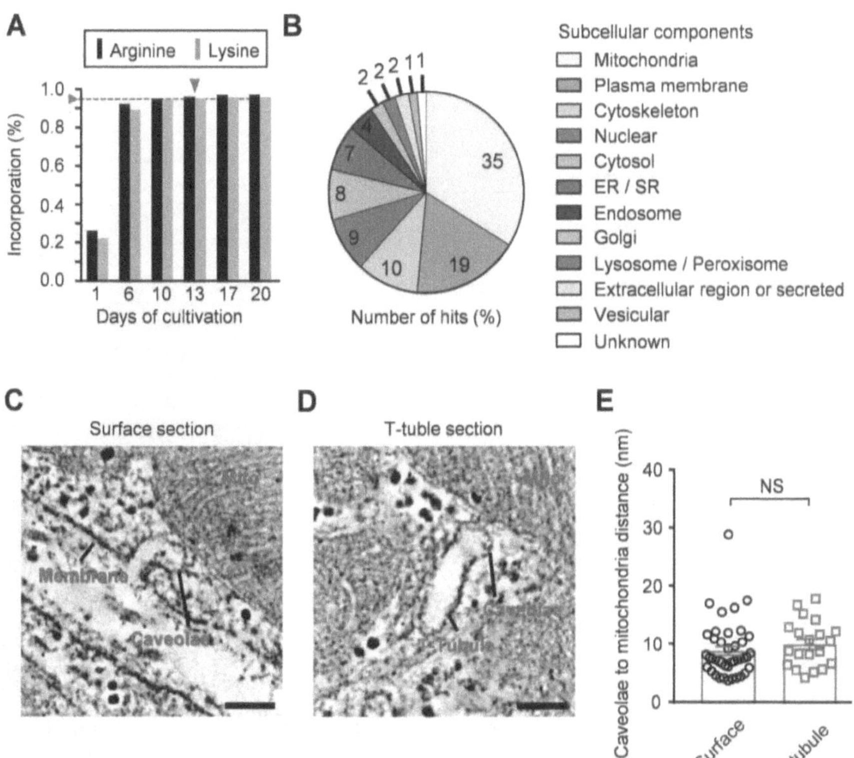


Supplement Figure 3.10 **SILAC incorporation, GO term analysis of V5-APEX2-CAV3 enriched proteins, and Electron Tomography. A**, Mass spectrometry (LC-MS/MS) analysis of heavy isotope-labeled ($^{13}C_6$, $^{15}N_4$-Arg and $^{13}C_6$, $^{15}N_2$-Lys), trypsin-digested NRCM cell lysates. SILAC incorporation reached ≥96% after 13 days of culture (red arrow). n=1. **B**, Subcellular component classification of V5-APEX2-CAV3 enriched proteins based on Gene Ontology (GO) annotation. The number of identified hits and there percent contribution are shown. **C-D**, Electron tomography images showing caveolae bulbs at the surface membrane (**C**) or at the transverse (T-) tuble (**D**) in nanometric proximity to mitochondria. Surface section n=41, T-tubule n=20. **E**, Bar graph summarizing the caveolae to mitochondria distance at the membrane surface (8.8 nm) and at T-tuble (9.9 nm). Student's t-test.

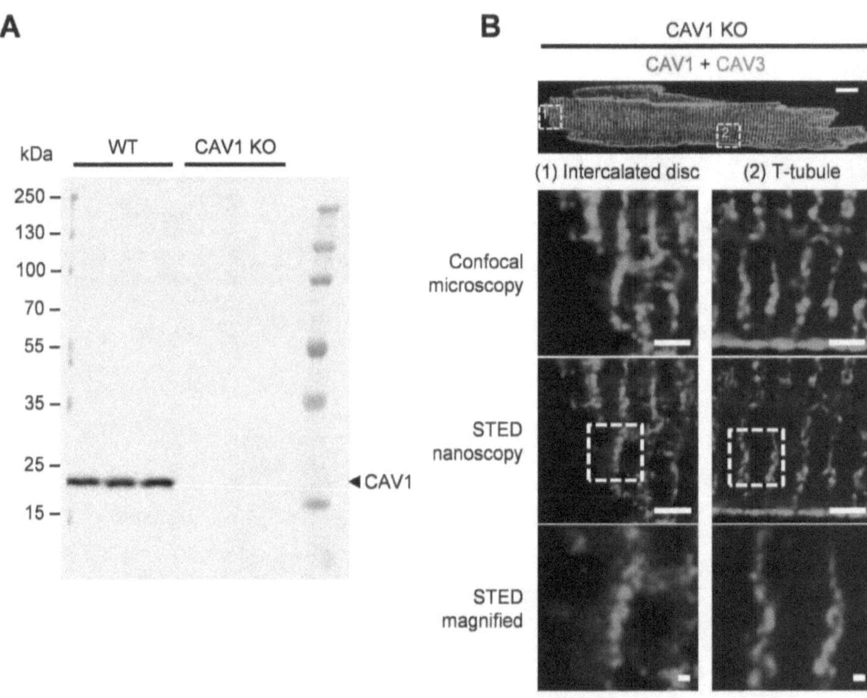



**Supplement Figure 3.11 CAV1 expression in ventricular cardiomyocytes.**
**A**, Immunoblot (full gel) showing singular CAV1 bands at 25 kDa in ventricular cardiomyocytes isolated from wild-type mouse hearts. CAV1 knockout mouse hearts were used to confirm specificity. n=3. **B,** Confocal and STED co-immunofluorescence imaging of CAV1 (red) and CAV3 (green) in a ventricular myocyte from a CAV1 knockout mouse heart. Of note, CAV1 staining showed unspecific signals at the intercalated disc, while no CAV1 signals were observed at the transverse (T-) tubules. Dashed boxes indicate magnified regions representing the intercalated disk and T-tubules. Scale bars: top 10 µm; Confocal microscopy 2 µm; STED nanoscopy 2 µm; STED magnified 200 nm.

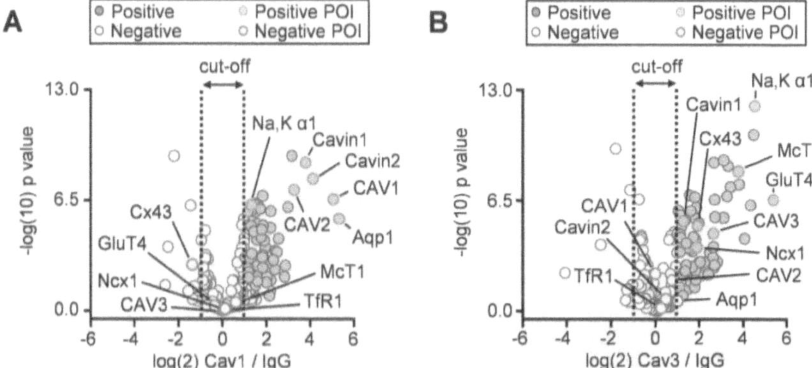


**Supplement Figure 3.12 CAV1 versus CAV3 protein interactions identified by AP-MS. A-B**, Volcano plots summarizing affinity-enriched CAV1 (**A**) and CAV3 (**B**) interactors identified by AP-MS. Significantly enriched proteins were identified by permutation-based false-discovery rate analysis (t-test, FDR=5%, S0=0.1) and logarithmic cut-off >1. n=3. Positive hits (blue circles) including functionally relevant proteins of interest (yellow circles) versus negative hits (open circles) and negative POI (red circles) as indicated by the legend.

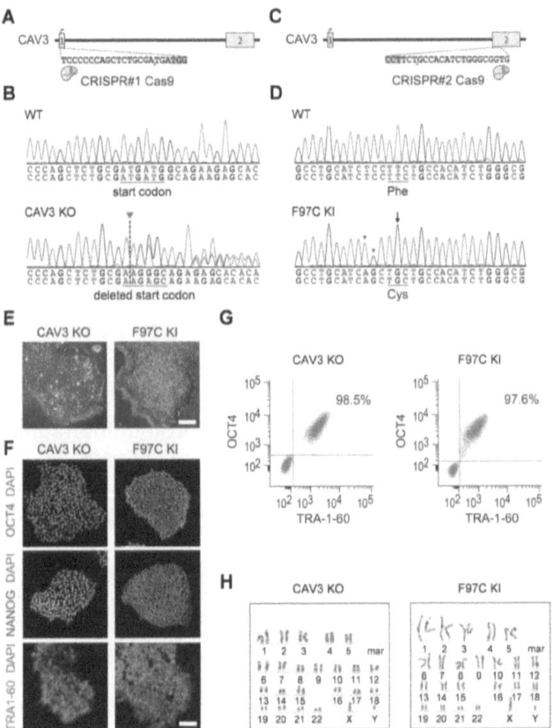



**Supplement Figure 3.13 CRISPR/Cas9 mediated CAV3 knock-out and F97C CAV3 knock-in. A**, CAV3 knockout iPSCs were generated with a CRISPR guide RNA targeting the start codon of the CAV3 gene and a clone with a start codon destruction on both alleles was selected for further analysis. **B**, Sanger sequencing of genomic DNA confirmed the deleted start codon in CAV3 knockout iPSCs. **C**, The F97C-CAV3 variant was introduced by CRISPR/Cas9-based homology-directed repair and a clone with a homozygous insertion was selected for further analysis. **D**, Sanger sequencing of genomic DNA confirmed the introduced SNP c.290T>G/p.F97C (arrow) in F97C iPSCs. Silent SNPs (asterisks) were introduced for mutation of the PAM site. E, Bright field imaging of the CAV3 KO and F97C KI iPSC lines exhibited a typical human stem cell-like morphology and proliferation characteristics. Scale bar: 100 µm. **F**, Immunofluorescence staining of the key pluripotency markers OCT4, NANOG and TRA1-60 in the CAV3 KO and F97C KI iPSC lines. Nuclei were counter-stained with DAPI. Scale bar: 100 µm. **G,** Purity of edited iPSC lines was evaluated by flow cytometry analysis of pluripotency markers OCT4 and TRA1-60. Gray dots represent the negative controls. **H**, Karyotypes of the edited iPSC lines between passages 25-30 demonstrated chromosomal stability after CRISPR/Cas9-based genomic editing and passaging.

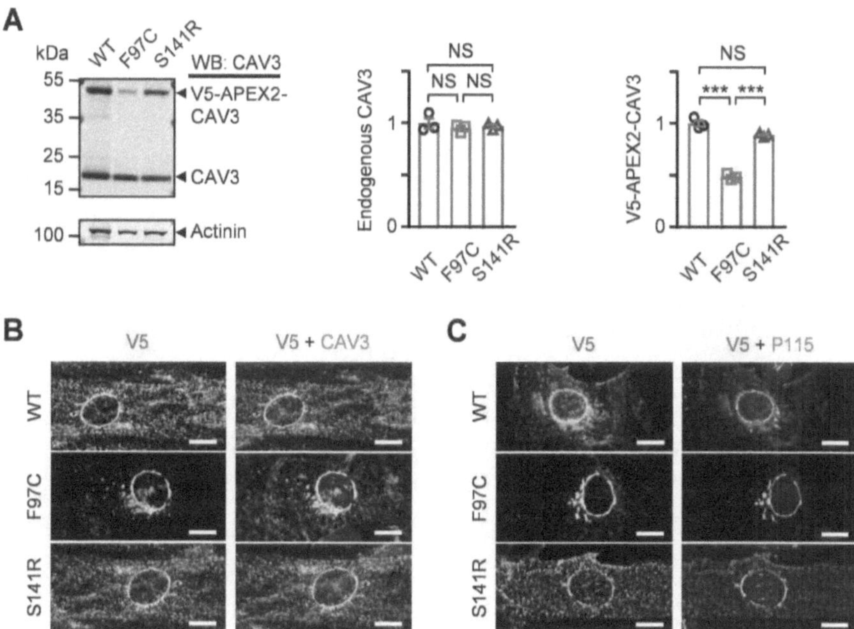



**Supplement Figure 3.14 Western blot and imaging of F97C and S141R V5-APEX2-CAV3. A**, Anti-CAV3 immunoblot detecting the expression of the V5-APEX2-CAV3 fusion proteins WT (control), F97C, and S141R in NRCM lysates. Bar graphs summarize the expression of endogenous versus recombinant CAV3 proteins each normalized to α-Actinin. n=3; ***p<0.001, Student's t-test. **B**, Confocal V5-immunofluorescence images showing a perinuclear and scattered accumulation of F97C V5-APEX2-CAV3. In contrast, WT and S141R mutant V5-APEX2-CAV3 showed peripherally distributed signals and co-localized with endogenous CAV3. Scale bars 10 μm. **B**, Confocal imaging of V5 and the Golgi marker P115. F97C mutant V5-APEX2-CAV3 accumulated in the trans-Golgi compartment. Nucleus stained with DAPI. Scale bars 10 μm.

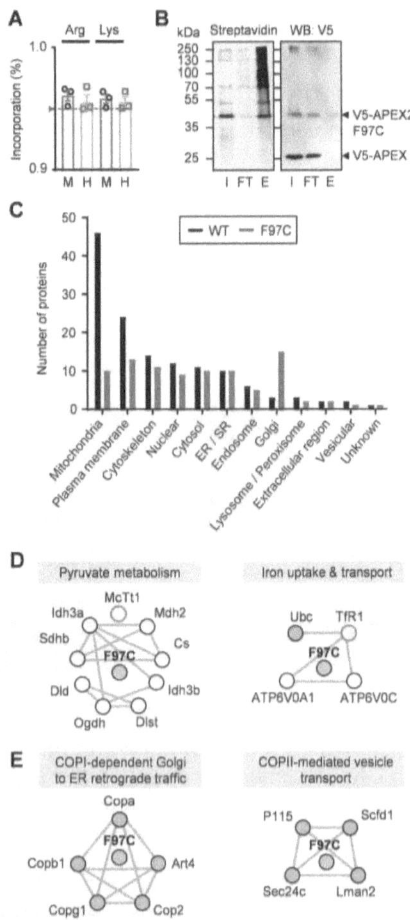


**Supplement Figure 3.15 Proximity proteomic analysis using V5-APEX2-CAV3-F97C. A**, LC-MS/MS detected ≥95% $^{13}C_6$,$^{15}N_4$-Arg and $^{13}C_6$,$^{15}N_2$-Lys incorporation in NRCM adenovirally transfected with V5-APEX2-CAV3-F97C (F97C). n=3. **B**, APEX2 biotinylated proteins were captured by streptavidin. I, input; FT, flow through; E, eluate. F97C and V5-APEX2 expression confirmed by V5 immunoblotting for input and flow through frations. n=3. **C**, Bar graph comparing the numerical changes in proteomic organelle components between WT and F97C positive hits. Note the strong increase in Golgi proteins. **D**, Analysis of identified proteins by the GO term 'pyruvate metabolism' based on the STRING database. The open circles indicate a loss of proximity; the red circle highlights the loss of McT1 and TfR1 proximity. Blue circles indicate a remaining interaction. Grey lines indicate protein interactions with a confidence score >0.7. **E**, STRING analysis confirming exclusive F97C-induced aberrant Golgi interactions for the indicated GO terms not present in WT. Grey lines indicate protein interactions with a confidence score >0.7.

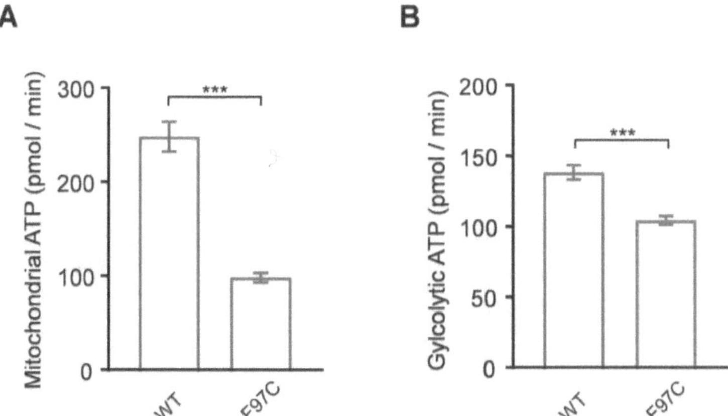



**Supplement Figure 3.16 Mitochondrial and glycolytic ATP in F97C CAV3 knock-in iPSC-CM. A-B**, Bar graphs compared basal mitochondrial (**A**) and glycolytic ATP (**B**) levels for WT versus F97C human cardiomyocytes. ATP-levels were calculated according to Agilent manufacture protocols. n=38; ***p<0.001, Student's t-test.

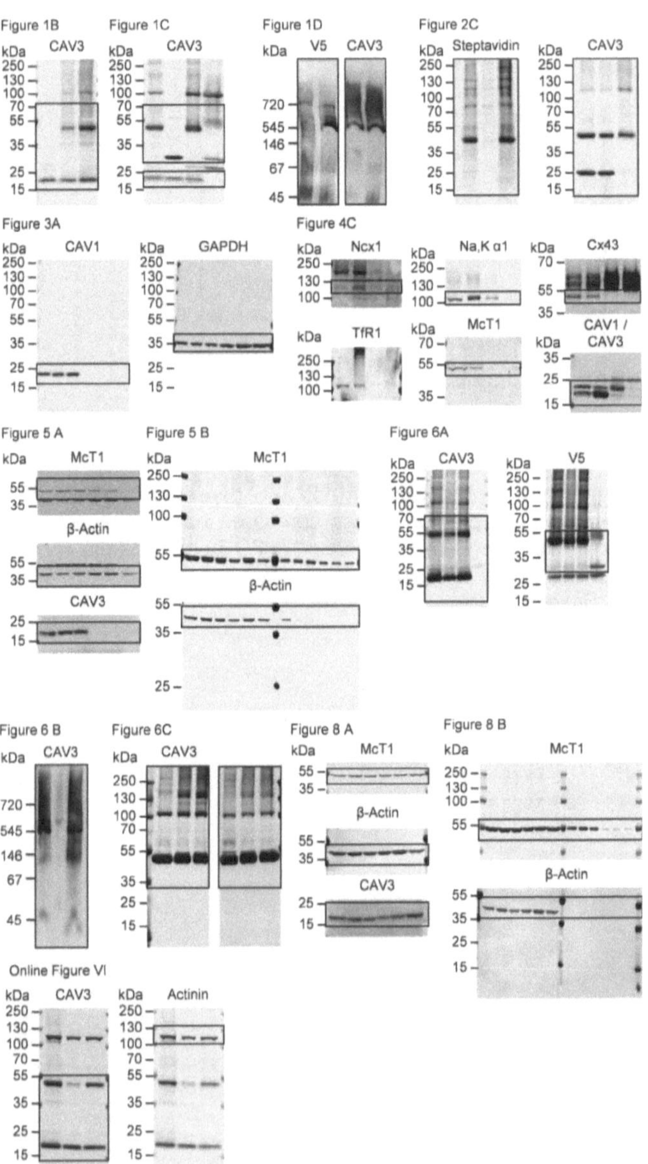

**Supplement Figure 3.17. Documentation of full scans of Western blots.** Frames indicate the data presented as manuscript figures or supplemental information.

## Supplemental tables and supporting information

**Supplement Table 3.1**  **ExaC based analysis of human CAV3 mutations**

| Human CA3 mutation[22] | Electrocardiogram QTc > 440 ms[22] | Allele frequency ExAc Rare frequency cut off < 0.0001 %[23] |
|---|---|---|
| G56S | - | > 0.010 % |
| C72W | - | > 0.001 % |
| T78M | 456 | > 0.003 % |
| A85T | - | - |
| F97C | 532 | - |
| S141R | 480 | - |

**Supplement Table 3.2**  **NRCM isolation buffer and cultivation medium**

**CBFHH buffer**

|  | MW(g/mol) | Final concentration |
|---|---|---|
| NaCl | 58.44 | 37 mM |
| KCl | 74.56 | 5.4 mM |
| KH2PO4 | 136.09 | 0.44 mM |
| Na2HPO4· 2 H2O | 177.99 | 33.5 mM |
| Glucose | 180.16 | 5.6 mM |
| HEPES | 238.31 | 20 mM |
| $MgSO_4$ | 120.37 | 0.8 mM |
| In 500 mL ddH$_2$O, pH 7.4 | | |

**NRCM cultivation medium**

|  | MW(g/mol) | Final concentration |
|---|---|---|
| FBS | - | 10 % (v/v) |
| 5-Bromo-2'-deoxyuridine | 307.10 | 10 mM |
| Penicillin/streptomycin | 647 | 1 % (v/v) |
| In 500 mL cell culture medium (DMEM-1 g/L D-glucose, Thermo Fisher Scientific) | | |

**Supplement Table 3.3**     **APEX2 biotinylation buffers**

**Quenching buffer**

|  | MW(g/mol) | Final concentration |
|---|---|---|
| 6-Hydroxy-2,5,7,8-tetramethylchroman-2-carboxylic-acid (Trolox) | 250.29 | 5 mM |
| Sodium azide | 65 | 10 mM |
| Sodium ascorbate | 136.09 | 10 mM |
| In 50 mL PBS (PBS, pH 7.4, without $Ca^{2+}$ and $Mg^{2+}$, Gibco) | | |

**RIPA quenching buffer**

|  | MW(g/mol) | Final concentration |
|---|---|---|
| Tris HCl, pH 7.4 | 157.60 | 50 mM |
| NaCl | 58.44 | 150 mM |
| Triton-X-100 | 647 | 1 % (v/v) |
| Sodium deoxycholate | 414.55 | 0.5 % (w/v) |
| Sodium dodecyl sulfate | 288.37 | 0.2 % (v/v) |
| 6-Hydroxy-2,5,7,8-tetramethylchroman-2-carboxylic-acid (Trolox) | 250.29 | 5 mM |
| Sodium azide | 65 | 10 mM |
| Sodium ascorbate | 136.09 | 10 mM |
| In 10 mL $ddH_2O$, pH 7.4 + 1 tablet of protease inhibitors (Complete Mini EDTA free, Sigma Aldrich) | | |

**Tris/HCl buffer containing urea**

|  | MW(g/mol) | Final concentration |
|---|---|---|
| Tris HCl, pH 8 | 157.60 | 50 mM |
| Urea | 60.06 | 2 mM |
| In 1 L $ddH_2O$, pH 8 | | |

**Biotin buffer**

|  | MW(g/mol) | Final concentration |
|---|---|---|
| Biotin | 244.31 | 2 mM |
| Sodium dodecyl sulfate | 288.37 | 2 % (v/v) |
| In 1 mL $ddH_2O$, pH 8 | | |

**Supplement Table 3.4**     **Mass spectrometry loading buffer**

**MS loading buffer**

|  | MW(g/mol) | Final concentration |
|---|---|---|
| Acetonitrile | 41.05 | 2 % (w/v) |
| Formic acid | 46.03 | 0.1 % (w/v) |
| In 50 mL $ddH_2O$ | | |

**Supplement Table 3.5** **Stem cell differentiation and human cardiomyocyte culture media**

| **iPSC culture medium** | | |
| --- | --- | --- |
| | MW(g/mol) | Final concentration |
| Thiazovivin | 311.4 | 2 µM |
| In 500 mL cell culture medium (StemMACS iPS-Brew stem cell culture media, Miltenyi Biotec) | | |

| **Cardio differentiation medium** | | |
| --- | --- | --- |
| | MW(g/mol) | Final concentration |
| Human recombinant albumin | - | 0.5 mg/mL |
| L-ascorbic acid 2-phosphate | 289.54 | 0.2 mg/mL |
| In 500 mL cell culture medium (RPMI 1640 cell culture medium with Glutamax and HEPES, Thermo Fisher Scientific ) | | |

| **Cardio selection medium** | | |
| --- | --- | --- |
| | MW(g/mol) | Final concentration |
| Human recombinant albumin | - | 0.5 mg/mL |
| L-ascorbic acid 2-phosphate | 289.54 | 0.2 mg/mL |
| lactate | 89.07 | 4 mM |
| In 500 mL cell culture medium (RPMI 1640 cell culture medium, no glucose, Thermo Fisher Scientific) | | |

| **Cardio culture medium** | | |
| --- | --- | --- |
| | MW(g/mol) | Final concentration |
| B27 | - | 2 % (v/v) |
| In 500 mL cell culture medium (RPMI 1640 cell culture medium with Glutamax and HEPES, Thermo Fisher Scientific ) | | |

Supplement Table 3.6    **Stem cell blocking buffer**

**Stem cell blocking buffer (immunofluorescence)**

| | MW(g/mol) | Final concentration |
|---|---|---|
| Bovine Serum Albumin | - | 1 % (v/v) |
| In 50 mL PBS (PBS, pH 7.4, without $Ca^{2+}$ and $Mg^{2+}$, Gibco) | | |

Supplement Table 3.7    **Cell lysis and protein transfer buffers**

**RIPA buffer**

| | MW(g/mol) | Final concentration |
|---|---|---|
| Tris HCl, pH 7.4 | 157.60 | 50 mM |
| NaCl | 58.44 | 150 mM |
| Triton-X-100 | 647 | 1 % (v/v) |
| Sodium deoxycholate | 414.55 | 0.5 % (w/v) |
| Sodium dodecyl sulfate | 288.37 | 0.2 % (v/v) |
| In 10 mL $ddH_2O$, pH 7.4 + 1 tablet of protease inhibitors (Complete Mini EDTA free, Sigma Aldrich) | | |

**Transfer buffer (immunoblot)**

| | MW(g/mol) | Final concentration |
|---|---|---|
| Tris HCl, pH 7.4 | 157.60 | 191.6 mM |
| Glycine | 75.07 | 1.92 M |
| In 1 L $ddH_2O$, pH 7.4 | | |

**5 x SDS buffer (immunoblot)**

| | MW(g/mol) | Final concentration |
|---|---|---|
| Tris HCl, pH 7.4 | 157.60 | 191.6 mM |
| Glycine | 75.07 | 1.92 M |
| SDS | 288.37 | 34.69 mM |
| β-Mercapto Ethanol | 78.31 | 5 % (v/v) |
| In 1 L $ddH_2O$, pH 8.3; dilute 2 mL 5 x SDS buffer in 3 mL to yield 2 x SDS buffer | | |

Supplement Table 3.8    **Seahorse XF assay buffer**

**Seahorse XF assay buffer**

| | MW(g/mol) | Final concentration |
|---|---|---|
| Pyruvate | 88.06 | 1 mM |
| Glucose | 180.16 | 4.5 mg/mL |
| In 500 mL Seahorse assay medium (Seahorse XF assay medium, Agilent) | | |

**Supplement Table 3.9    HEK293A cell culture medium**

**HEK293A cell culture medium**

|  | MW(g/mol) | Final concentration |
|---|---|---|
| FBS | - | 10 % (v/v) |
| L-glutamine | 146.14 | 2 mM |
| penicillin/streptomycin | - | 1 % (v/v) |
| In 500 mL cell culture medium (Dulbecco's Modified Eagle's Medium - low glucose, Sigma Aldrich) | | |

**Supplement Table 3.10    Blue Native (BN)-PAGE buffers**

**Homogenization buffer**

|  | MW(g/mol) | Final concentration |
|---|---|---|
| Sucrose | 342.30 | 250 mM |
| Tris HCl, pH 7.4 | 157.60 | 10 mM |
| EDTA | 292.24 | 1 mM |
| PMSF | 174.19 | 1 mM |
| In 50 mL ddH$_2$O, pH 7.4 | | |

**Solubilization buffer**

|  | MW(g/mol) | Final concentration |
|---|---|---|
| NaCl | 58.44 | 50 mM |
| imidazole | 68.08 | 50 mM |
| EDTA | 292.24 | 1 mM |
| aminocaproic acid | 414.55 | 2 mM |
| In 50 mL ddH$_2$O, pH 7.4 | | |

**Supplement Table 3.11 Mouse cardiomyocyte isolation, blocking, and permeabilization buffers**

### Perfusion buffer (mouse heart)

|  | MW(g/mol) | Final concentration |
|---|---|---|
| NaCl | 58.44 | 120.4 mM |
| KCl | 74.56 | 14.7 mM |
| KH2PO41 | 36.09 | 0.6 mM |
| Na2HPO4· 2 H2O | 177.99 | 0.6 mM |
| MgSO4· 7 H2O | 246.48 | 1.2 mM |
| HEPES | 238.31 | 10 mM |
| NaHCO3 | 84.01 | 4.6 mM |
| Taurin | 125.20 | 30 mM |
| 2,3-Butanedione monoxime | 101.1 | 10 mM |
| Glucose | 180.16 | 5.5 mM |
| In 1 L ddH$_2$O, pH 7.4 |  |  |

### Digestion buffer (mouse heart)

|  | MW(g/mol) | Final concentration |
|---|---|---|
| Collagenase type II | - | 2 mg/mL |
| CaCl$_2$ | 110.98 | 40 µM |
| In 50 mL perfusion buffer, pH 7.4 |  |  |

### Stop buffer (mouse heart)

|  | MW(g/mol) | Final concentration |
|---|---|---|
| Bovine calf serum | - | 10 % (v/v) |
| CaCl2 | 110.98 | 12.5 µM |
| In 50 mL perfusion buffer, pH 7.4 |  |  |

### Blocking/permeabilization buffer (mouse cardiomyocytes immunofluorescence)

|  | MW(g/mol) | Final concentration(mM) |
|---|---|---|
| Bovine calf serum | - | 10 % (v/v) |
| Triton X-100 | 74.56 | 0.2 % (v/v) |
| In 50 mL PBS (PBS, pH 7.4, without Ca$^{2+}$ and Mg$^{2+}$, Gibco) |  |  |

**Supplement Table 3.12 Electron tomography buffer**

### Reynold's lead citrate

|  | MW(g/mol) | Final concentration |
|---|---|---|
| Pb(NO$_3$)$_2$ | 331.23 | 80 mM |
| Na$_3$C$_6$H$_5$O$_7$ | 258.06 | 136 mM |
| NaOH, 1N | 40.00 | 8 mL (v/v) |
| In 50 mL ddH$_2$O, pH 12 |  |  |

**Supplement Table 3.13   Immunoprecipitation buffers**

**CHAPS co-IP buffer**

|  | MW(g/mol) | Final concentration |
|---|---|---|
| Tris HCl, pH 7.4 | 157.60 | 50 mM |
| NaCl | 58.44 | 150 mM |
| CHAPS | 614.88 | 0.15 % (w/v) |
| EGTA | 380.35 | 1 mM |

In 10 mL ddH$_2$O, pH 7.4 + 1 tablet of protease inhibitors (Complete Mini EDTA free, Sigma Aldrich)

**Sodium deoxycholate co-IP buffer**

|  | MW(g/mol) | Final concentration |
|---|---|---|
| Tris HCl, pH 7.4 | 157.60 | 50 mM |
| NaCl | 58.44 | 150 mM |
| Triton-X-100 | 614.88 | 1 % (v/v) |
| Sodium deoxycholat | 414.55 | 0.5 % (w/v) |

In 10 mL ddH$_2$O, pH 7.4 + 1 tablet of protease inhibitors (Complete Mini EDTA free, Sigma Aldrich)

**Supplemental references**


1.  Rutering J, Ilmer M, Recio A, Coleman M, Vykoukal J, Alt E. Improved Method for Isolation of Neonatal Rat Cardiomyocytes with Increased Yield of C-Kit+ Cardiac Progenitor Cells. *J Stem Cell Res Ther*. 2015;5:1–8.

2.  Lam SS, Martell JD, Kamer KJ, Deerinck TJ, Ellisman MH, Mootha VK, Ting AY. Directed evolution of APEX2 for electron microscopy and proximity labeling. *Nat Methods*. 2015;12:51–54.

3.  Hung V, Zou P, Rhee H-W, Udeshi ND, Cracan V, Svinkina T, Carr SA, Mootha VK, Ting AY. Proteomic Mapping of the Human Mitochondrial Intermembrane Space in Live Cells via Ratiometric APEX Tagging. *Molecular Cell*. 2014;55:332–341.

4.  Callegari S, Müller T, Schulz C, Lenz C, Jans DC, Wissel M, Opazo F, Rizzoli SO, Jakobs S, Urlaub H, Rehling P, Deckers M. A MICOS–TIM22 Association Promotes Carrier Import into Human Mitochondria. ScienceDirect. 2019, 10.1016.

5.  Atanassov I, Urlaub H. Increased proteome coverage by combining PAGE and peptide isoelectric focusing: Comparative study of gel-based separation approaches. *Proteomics*. 2013;13:2947–2955.

6.  Lambert J-P, Ivosev G, Couzens AL, Larsen B, Taipale M, Lin Z-Y, Zhong Q, Lindquist S, Vidal M, Aebersold R, Pawson T, Bonner R, Tate S, Gingras A-C. Mapping differential interactomes by affinity purification coupled with data-independent mass spectrometry acquisition. *Nature Methods*. 2013;10:1239–1245.

7.  Zhang Y, Bilbao A, Bruderer T, Luban J, Strambio-De-Castillia C, Lisacek F, Hopfgartner G, Varesio E. The Use of Variable Q1 Isolation Windows Improves Selectivity in LC-SWATH-MS Acquisition. *J Proteome Res*. 2015;14:4359–4371.

8.  Coy-Vergara J, Rivera-Monroy J, Urlaub H, Lenz C, Schwappach B. A trap mutant reveals the physiological client spectrum of TRC40. Journal of Cell Science, 2019, 10.1242.

9.  Erdmann J, Thöming JG, Pohl S, Pich A, Lenz C, Häussler S. The Core Proteome of Biofilm-Grown Clinical Pseudomonas aeruginosa Isolates. *Cells*. 2019;8:1129.

10.  Szklarczyk D, Gable AL, Lyon D, Junge A, Wyder S, Huerta-Cepas J, Simonovic M, Doncheva NT, Morris JH, Bork P, Jensen LJ, Mering C von. STRING v11: protein–protein association networks with increased coverage, supporting functional discovery in genome-wide experimental datasets. *Nucleic Acids Research*. 2019;47:D607–D613.

11.  Oliveira GS, Santos AR. Using the Gene Ontology tool to produce de novo protein-protein interaction networks with IS_A relationship. *Genet Mol Res*. 2016;15.

12.  Cyganek L, Tiburcy M, Sekeres K, Gerstenberg K, Bohnenberger H, Lenz C, Henze S, Stauske M, Salinas G, Zimmermann W-H, Hasenfuss G, Guan K. Deep phenotyping of human induced pluripotent stem cell–derived atrial and ventricular cardiomyocytes. *JCI Insight*. 2018;3.

13.  Zhao L, Hayes K, Glassman A. A simple efficient method of sequential G-banding and fluorescence in situ hybridization. *Cancer Genet Cytogenet*. 1998;103:62–64.

14.  El-Battrawy I, Müller J, Zhao Z, Cyganek L, Zhong R, Zhang F, Kleinsorg M, Lan H, Li X, Xu Q, Huang M, Liao Z, Moscu-Gregor A, Albers S, Dinkel H, Lang S, Diecke S, Zimmermann WH, Utikal J, Wieland T, Borggrefe M, Zhou X, Akin I. Studying Brugada Syndrome With an SCN1B Variants in Human-Induced Pluripotent Stem Cell-Derived Cardiomyocytes. Frontiers in Cell and Developmental Biology. 2019; 10.3389.

15.  Dudek J, Cheng I, Chowdhury A, Wozny K, Balleininger M, Reinhold R, Grunau S, Callegari S, Toischer K, Wanders RJ, Hasenfuß G, Brügger B, Guan K, Rehling P. Cardiac-specific succinate dehydrogenase deficiency in Barth syndrome. *EMBO Mol Med*. 2016;8:139–154.

16. O'Connell TD, Rodrigo MC, Simpson PC. Isolation and culture of adult mouse cardiac myocytes. *Methods Mol Biol*. 2007;357:271–296.

17. Wagner E, Brandenburg S, Kohl T, Lehnart SE. Analysis of Tubular Membrane Networks in Cardiac Myocytes from Atria and Ventricles. *J Vis Exp*. 2014;

18. Brandenburg S, Pawlowitz J, Fakuade FE, Kownatzki-Danger D, Kohl T, Mitronova GY, Scardigli M, Neef J, Schmidt C, Wiedmann F, Pavone FS, Sacconi L, Kutschka I, Sossalla S, Moser T, Voigt N, Lehnart SE. Axial Tubule Junctions Activate Atrial Ca2+ Release Across Species. *Front Physiol*. 2018;9.

19. Mastronarde DN. Dual-Axis Tomography: An Approach with Alignment Methods That Preserve Resolution. *Journal of Structural Biology*. 1997;120:343–352.

20. Mastronarde D. Tomographic Reconstruction with the IMOD Software Package. *Microsc Microanal*. 2006;12:178–179.

21. Kremer JR, Mastronarde DN, McIntosh JR. Computer visualization of three-dimensional image data using IMOD. *J Struct Biol*. 1996;116:71–76.

22. Vatta M, Ackerman MJ, Ye B, Makielski JC, Ughanze EE, Taylor EW, Tester DJ, Balijepalli RC, Foell JD, Li Z, Kamp TJ, Towbin JA. Mutant Caveolin-3 Induces Persistent Late Sodium Current and Is Associated With Long-QT Syndrome. *Circulation*. 2006;114:2104–2112.

23. Whiffin N, Minikel E, Walsh R, O'Donnell-Luria AH, Karczewski K, Ing AY, Barton PJR, Funke B, Cook SA, MacArthur D, Ware JS. Using high-resolution variant frequencies to empower clinical genome interpretation. *Genet Med*. 2017;19:1151–1158.

# 4 Additional Methods

Methods that are described in the manuscript are not stated again. All buffers that are not listed in this additional method part are specified in the buffer lists (Supplement Tables 3.2-3.13) in the supplemental method part of the manuscript.

Used adenoviruses, cell lines, chemicals, drugs, kits, cell-culture media, consumables, general equipment and software are listed in the Appendix Table 8.1–8.9 and the used antibodies with detailed information are listed in Appendix Table 8.10.

## 4.1 Co-immunoprecipitation of CAV3 interacting proteins

Mouse ventricles were prepared from perfused mice hearts as described in the supplemental methods of the manuscript (**Adult mouse ventricular cardiomyocyte isolation**). Ventricular tissues from nine adult mice (12 weeks) were manually dissected under a binocular microscope (Stemi 305, Zeiss), snap-frozen in liquid N2, cut into small pieces, and stored at -80 °C until further use. Next, ventricular tissues of three mice, were solubilized with either 1 mL CHAPS-, 1 mL sodium deoxycholate- or 1 mL octylglucoside co-IP buffer, respectively (Figure 5.1 A; Table 4.1). The protein concentration was determined by absorption measurement (Pierce BCA Protein Assay Kit; Thermo Fisher Scientific), and lysates of 500 µg were incubated with 3 µg anti-CAV3 antibody (ab2912, Abcam) or normal rabbit IgG (12-370, Merck) antibody at 4 °C overnight in an overhead rotator (Overhead rotator, Bio Grant). Magnetic beads (Dynabeads Protein G, Thermo Fisher Scientific) were equilibrated in a ratio 1:1 in the respective co-IP buffer and 30 µL equilibrated magnetic beads were added to the samples and incubated for 2 h at 4 °C (Overhead rotator, Bio Grant). The magnetic beads were extracted with a magnet (DynaMag-2 Magnet, Thermo Fisher Scientific), the solution was discarded and the beads were washed thrice with ice-cold 500 µL of the respective co-IP buffer to minimize unspecific binding. Precipitated proteins were eluted in 60 µL of 2 × SDS buffer containing β-mercaptoethanol (Supplement Table 3.7). The co-IP samples were analyzed by label-free SWATH-MS (Sequential Window Acquisition of All THeoretical Mass Spectra) as described in the supplemental methods of the manuscript (**Sample preparation for label-free SWATH-MS (Sequential Window Acquisition of All THeoretical Mass Spectra) and NanoLC-MS/MS analysis by label-free SWATH-MS)**. Mass spectrometry was carried out by the proteomic service unit (Dr. Christof Lenz, Institute of Clinical Chemistry, University Medical Center Göttingen, Göttingen).

**Table 4.1 Immunoprecipitation buffers**

**CHAPS co-IP buffer**

|  | MW(g/mol) | Final concentration |
|---|---|---|
| Tris HCl, pH 7.4 | 157.60 | 50 mM |
| NaCl | 58.44 | 150 mM |
| CHAPS | 614.88 | 0.15 % (w/v) |
| EGTA | 380.35 | 1 mM |

In 10 mL ddH$_2$O, pH 7.4 + 1 tablet of protease inhibitors (Complete Mini EDTA free, Sigma Aldrich)

**Sodium deoxycholate co-IP buffer**

|  | MW(g/mol) | Final concentration |
|---|---|---|
| Tris HCl, pH 7.4 | 157.60 | 50 mM |
| NaCl | 58.44 | 150 mM |
| Triton-X-100 | 614.88 | 1 % (v/v) |
| Sodium deoxycholat | 414.55 | 0.5 % (w/v) |

In 10 mL ddH$_2$O, pH 7.4 + 1 tablet of protease inhibitors (Complete Mini EDTA free, Sigma Aldrich)

**Octylglucoside co-IP buffer**

|  | MW(g/mol) | Final concentration |
|---|---|---|
| Tris HCl, pH 7.4 | 157.60 | 50 mM |
| NaCl | 58.44 | 150 mM |
| Triton-X-100 | 614.88 | 1 % (v/v) |
| Octyl β-D-glucopyranoside | 292.37 | 60 mM |

In 10 mL ddH$_2$O, pH 7.4 + 1 tablet of protease inhibitors (Complete Mini EDTA free, Sigma Aldrich)

## 4.2 Adult mouse atria cardiomyocyte isolation

Similar to the protocol described for ventricular cardiomyocytes, adult atrial cardiomyocytes were perfused and digested as described in the supplemental methods of the manuscript (**Adult mouse ventricular cardiomyocyte isolation**). Atria tissue was dissected under a binocular microscope (Stemi 305, Zeiss) using microsurgical scissors (Fine Science Tools GmbH, 15025-10) and gently minced in 1 mL digestion buffer (Supplement Table 3.11). Digestion was stopped by adding 1.5 mL stopping buffer (Supplement Table 3.11) and the suspension was transferred into a 15 mL Falcon tube (1048400, Sarstedt). Isolated atrial cardiomyocytes were centrifuged at 200 x g for 5 min (Heraeus, Fresco 21 centrifuge, Thermo Fisher Scientific) and resuspended in 2 mL perfusion buffer (Supplement Table 3.11). Cell quality was documented by transmitted light imaging (Zeiss LSM 710 and 880, Jena, Germany) using Fiji (https://imagej.net/Fiji) following criteria documented previously.[133]

## 4.3 Superresolution STED immunofluorescence nanoscopy

Isolated atrial and ventricular cardiomyocytes were prepared and fixed for STED microscopy as described in the supplemental methods of the manuscript (**Confocal microscopy and superresolution STED immunofluorescence nanoscopy**). The fixed samples were incubated overnight at 4 °C in 500 µL blocking/permeabilization buffer (Supplement Table 3.11) followed by an incubation overnight at 4 °C with TfR1 (13-6800, Invitrogen) and CAV3 (2912, Abcam) or CAV1 (Ab17052, Abcam) and CAV3 (2912, Abcam), diluted in blocking buffer to a final concentration of 1 µg antibody/mL in blocking/permeabilization buffer. After washing thrice with blocking/permeabilization buffer, samples were incubated with secondary antibodies diluted 1:1000 overnight at 4 °C. For super resolution STED microscopy, STAR 635P (2-0002-007-5, Abberior) and STAR 580 (2-0012-005-8) was used. After washing thrice with PBS (PBS, pH 7.4, without $Ca^{2+}$ and $Mg^{2+}$, Gibco), samples were embedded in DAPI-free mounting medium (ProLong Gold Antifade Mountant, Thermo Fisher Scientific). Embedded samples were stored overnight at RT and imaged the next day.

## 4.4 Transferrin and Cholesterol-PEG-KK114 live labeling in ventricular cardiomyocytes

For the live labeling of ventricular cardiomyocytes with differic Transfferin-488 (differic Transferrin Alexa Fluor conjugate 488, Thermo Fisher Scientific) and a custom-made photostable cholesterol dye (Cholesterol-PEG-KK114)[8], isolated ventricular cardiomyocytes were plated on glass coverslips (Ø 18 mm, width 1.5 mm, Menzel) after coating with laminin (2 mg/mL) at a dilution of 1:10 in perfusion buffer (Supplement Table 3.11). Living ventricular cardiomyocytes were incubated with 250 µg differic Transfferin-488 diluted in 500 µL PBS (PBS, pH 7.4, without $Ca^{2+}$ and $Mg^{2+}$, Gibco) for 5 min on ice, according to the manufacture protocol (Transferrin conjugates, Thermo Fischer Scientific). After washing thrice with ice cold PBS (PBS, pH 7.4, without $Ca^{2+}$ and $Mg^{2+}$, Gibco), living ventricular cardiomyocytes were incubated 15 min at RT with 250 nM Chol-PEG-KK114, according to our previous published protocol.[8] After incubation, living ventricular cardiomyocytes were washed thrice with ice cold PBS (PBS, pH 7.4, without $Ca^{2+}$ and $Mg^{2+}$, Gibco).

For STED nanoscopy a Leica TCS SP8 system with a HC PL APO C2S 100x/1.40 oil objective and a pixel size 16.23 x 16.23 nm was used. For live imaging, KK114 was exited at 635 nm, and fluorescence was detected at 650–700 nm. Differic Transfferin-488 was exited at 488 nm, and fluorescence was detected at 490–540 nm. For STED depletion of KK114 a 775 nm laser beam was used, and for STED depletion of Transfferin-488 a 595 nm laser beam was used. To maximize the imaging resolution, the STED laser power was adjusted following previously established workflows.[133] Raw images were processed in Fiji (https://imagej.net/Fiji) following established protocols.[133]

## 4.5 iPSC cardiomyocyte cell surface biotinylation of TfR1

Stem cell differentiation and genome editing was carried out by the stem cell service unit (Dr. Lukas Cyganek, Stem Cell Unit, Göttingen).
Human iPSC-WT or CAV3 knock-out derived cardiomyocytes were seeded at a density of 1 million cells on 35 mm dishes (CELLSTAR 6-well plate, Greiner) and cultured for 60 days. For cell surface biotinylation, iPSC-cardiomyocytes were washed thrice with 500 µL ice cold PBS (PBS, pH 7.4, without $Ca^{2+}$and $Mg^{2+}$, Gibco) and incubated for 1 h at 4 °C with 2 mM tagging solution (EZ-Link Sulfo-NHS-Biotin, Thermo Fisher Scientific) or PBS (PBS, pH 7.4, without $Ca^{2+}$ and $Mg^{2+}$, Gibco) as negative control. Further, sample preparation and avidin pulldown of biotinylated proteins were performed as described in supplemental methods of the manuscript (**iPSC-cardiomyocyte cell surface biotinylation and elution of biotinylated surface proteins**).

## 4.6 Immunoblotting of iPSC derived cardiomyocytes and isolated mouse cardiomyocytes

Stem cell differentiation and genome editing was carried out by the stem cell service unit (Dr. Lukas Cyganek, Stem Cell Unit, Göttingen).
For immunoblotting of iPSC derived and isolated cardiomyocytes, cells were homogenized and the proteins were transferred onto PVDF membranes as described in the supplemental methods of the manuscript (**Immunoblotting and streptavidin blotting for protein analysis**).
PVDF membranes were incubated with 1 µg / mL TfR1, CAV3, β-Actin or CAV1 (Table 8.10) in 5 % w/v non-fat milk (Milkpowder, Roth) in Tris-buffered saline with 0.05 % v/v Tween (Tween 20, Sigma Aldrich) at 4 °C overnight. Followed by washing thrice with PBS (pH 7.4, without $Ca^{2+}$ and $Mg^{2+}$, Gibco) and incubated with fluorescent anti-mouse or anti-rabbit secondary antibodies (P/N 926-32212, P/N 926-68072, P/N 926-32213, P/N 926-68073, IRDye LI-COR) at a dilution of 1:15,000 at 4 °C overnight. Fluorescence signals were captured

with the Odyssey CLx imaging system (LI-COR) and band intensities quantified with Image Studio Lite Version 5.2 (LI-COR).

## 4.7 Proximity analysis for CAV3-S141R biotinylated proteins

The APEX2 assay, avidin capture of biotinylated proteins, MS analysis and APEX2 data processing for V5-APEX2-CAV3-S141R was performed as described for the V5-APEX2-CAV3-F97C in the supplemental methods of the manuscript (**page 50-54**). Mass spectrometry was carried out by the proteomic service unit (Dr. Christof Lenz, Institute of Clinical Chemistry, University Medical Center Göttingen, Göttingen).
Following protein identification by mass spectrometry, we analyzed the GO terms 'caveolae', 'pyruvate metabolism', and 'iron uptake & transport' based on the STRING V11 database for protein–protein interaction networks (string-db.org). We used a scoring cut-off of ≥0.75 to define positive interactions following published workflows.[134]

## 4.8 Statistical analysis

Data are presented as mean ± standard error of the mean (SEM). Unpaired 2-tailed Student's t-test or 1-way-ANOVA was applied as specified in the figure legends. A p value of less than 0.05 was considered statistically significant. Statistical analyses were performed with Graph Pad Prism version 7.03.

# 5 Additional Results

## 5.1 Validation of CAV immunoprecipitation conditions

To validate the WT V5-APEX2-CAV3 proximity proteomic data, which include the monocarboxylate transporter 1 (McT1) and the transferrin receptor 1 (TfR1) as potentially new CAV3 interactors (manuscript Figure 3.4) and to further dissect CAV1 and CAV3 isoform specific protein interactions, we established immunoprecipitation protocols for CAV3 in mouse ventricular tissue lysates. For that purpose, three different lysis buffers were tested in parallel according to previously reported CAV3 immunoprecipitation protocols in heterologous cell systems.[135,136,123] Specifically, three whole ventricular tissues were lysed with either octyl β-D-glucopyranoside (OGP), sodium deoxycholate (DCA) or 3-((3-cholamidopropyl) dimethylammonio)-1-propanesulfonate (CHAPS), and CAV3 affinity purified interactors were quantified by Sequential Window Acquisition of All Theoretical Mass Spectra (AP/SWATH-MS) (Figure 5.1 A). Mass spectrometry was carried out by the proteomic service unit (Dr. Christof Lenz, Institute of Clinical Chemistry, University Medical Center Göttingen, Göttingen). Positive hits were identified by permutation-based false-discovery rate analysis (p<0.05)[137], and a log2 fold change of the CAV3 versus IgG ratio >1 was defined as cut-off criteria, according to previous established protocols.[138]

As compared to the control (IgG), we identified 12 potential CAV3 protein interactions using OGP, 17 potential interactions using DCA, and 70 potential interactions using CHAPS (Figure 5.1 B-D, Table 8.15-8.17). Importantly, proteins-of-interest (POIs) identified by proximity proteomic analysis (manuscript Figure 3.2), including the essential core complex caveolar proteins Cavin1 and Cavin4,[139,140] as well as McT1, Na,K-ATPase α1, Ncx1, Connexin43, and the insulin-dependent glucose transporter (GluT4) were comprehensively identified by CHAPS solubilisation and SWATH-MS (Figure 5.1 D). Therefore, CHAPS was used for the identification of isoform-specific CAV interactions (manuscript Figure 3.4 A-B). Finally, CAV3 immunoprecipitation experiments followed by immunoblotting confirmed McT1 and TfR1 as CAV3 interactors (manuscript Figure 3.4 C).

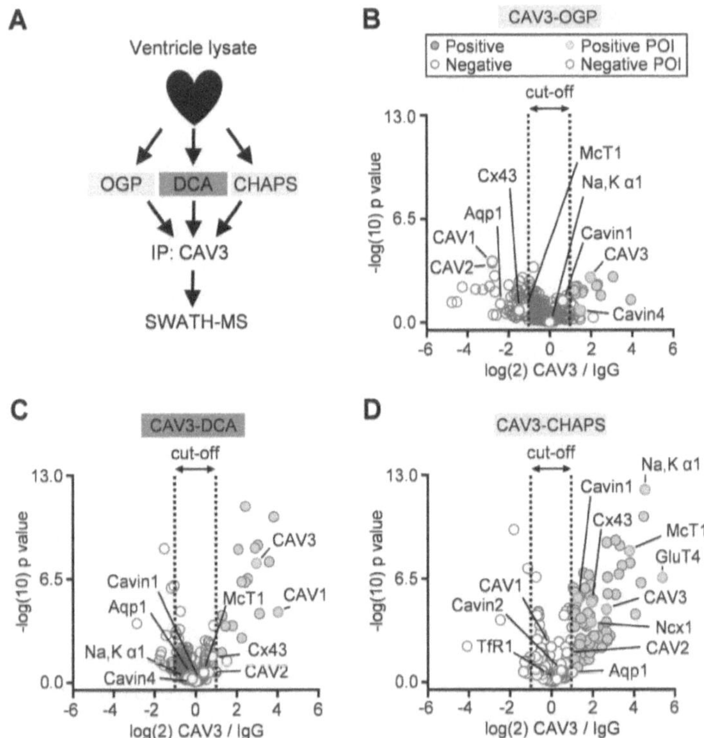



**Figure 5.1 Validation of CAV3 immunoprecipitation conditions by AP and SWATH-MS. A**, Workflow used for CAV3 immunoprecipitation in mouse ventricular tissue lysates. Three whole ventricular tissues were lysed with octyl β-D-glucopyranoside (OGP), sodium deoxycholate (DCA) or 3-((3-cholamidopropyl) dimethylammonio)-1-propanesulfonate (CHAPS) and 500 µg tissue lysate incubated with CAV3 antibody. CAV3 interactors were enriched by immunoprecipitation (IP) and analyzed by SWATH-MS. **B-D**, Volcano plots showing CAV3 enriched protein interactors for tissue lysates solubilized with OGP (**B**), DCA (**C**) or CHAPS (**D**). Logarithmic ratios identified enriched CAV3 interacting proteins as indicated: positive hits (blue circles) including functionally relevant proteins of interest (POIs, yellow circles). Positive hits and POIs were identified by permutation-based false-discovery rate analysis (t-test, p>0.05, FDR=5 %, S0=0.1) and a log2 fold change >1 defined as cut-off criteria (dashed line). Negative hits (red circles) were excluded based on the same criteria. n=3.

## 5.2 TfR1 and Transferrin are localized in close proximity to CAV3 nanodomains

Using affinity proteomics we confirmed TfR1 as a novel candidate CAV3 interactor (manuscript Figure 3.4 C), suggesting that CAV3 stabilizes TfR1 in the surface membrane as a prerequisite for Transferrin uptake. To investigate the relationship between CAV3 and TfR1 in mature adult mouse cardiomyocytes, we used confocal (LSM 710 system, Plan-Apochromat 63x/1.40 oil objective, Zeiss) and STimulated Emission Depletion (STED) microscopy (TCS SP8 system, HC PL APO C2S 100x/1.40 oil objective, Leica). While confocal imaging resulted in blurred TfR1 (13-6800, Invitrogen) signal blobs, STED nanoscopy resolved small punctate TfR1 cluster signals in the lateral surface membrane and in transverse (T-)tubules in close proximity to CAV3 (2912, Abcam) cluster signals (Figure 5.2 A).

Since TfR1 facilitates the binding and subsequent endocytosis of monoferric and differic Transferrin,[141] we hypothesized that iron-bounded Transferrin is located in immediate proximity to CAV3 clusters. To investigate the localization of iron-bounded Transferrin, we labeled living ventricular cardiomyocytes with extracellular differic Transferrin-Alexa488 (differic Transferrin Alexa Fluor conjugate 488, Thermo Fisher Scientific). Confocal imaging revealed differic-Transferrin cluster signals across the cell surface (Figure 5.2 B, *top*). Cardiomyocytes not exposed to differic Transferrin-Alexa488 (negative control) confirmed the specificity of differic Transferrin-Alexa488 signals (Figure 5.2 B, *bottom*).

To validate if the differic Transferrin signals are also in close proximity to cholesterol-rich CAV3 membrane nanodomains in living cardiomyocytes,[23] we used a custom-made photostable cholesterol dye (Cholesterol-PEG-KK114) for staining.[8] Co-labeling of living ventricular cardiomyocytes with differic Transferrin-Alexa488 and Cholesterol-PEG-KK114 indeed showed fluorescent differic Transferrin-Alexa488 signals adjacent to cholesterol membrane domains at the lateral surface and in T-tubules in cardiomyocytes (Figure 5.2 C). These data suggested that differic Transferrin binds to cholesterol-rich nanodomains stabilized by CAV3 clusters, namely caveolae and T-tubules.

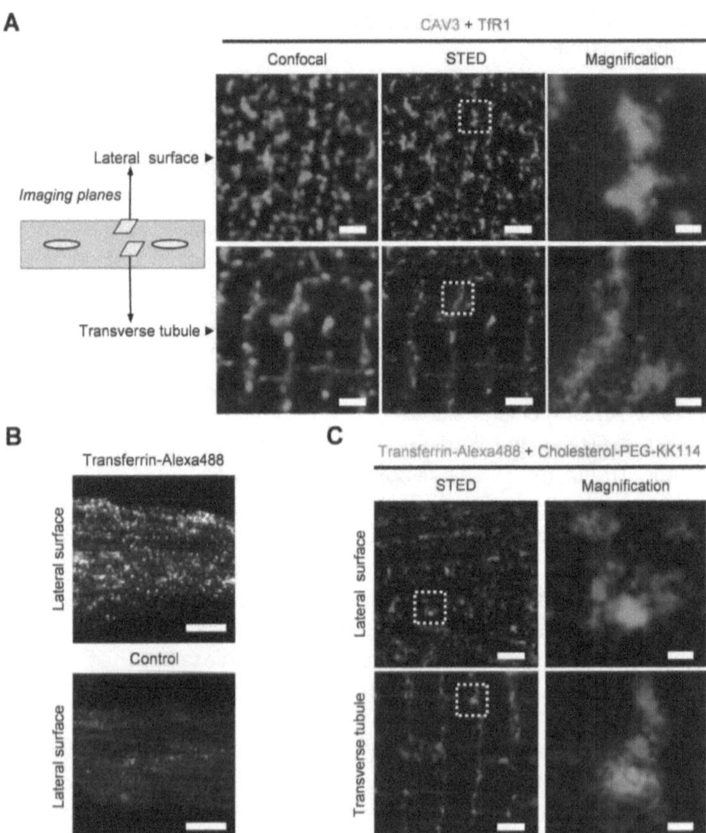



**Figure 5.2 Distribution of TfR1 and Transferrin in ventricular cardiomyocytes**. **A**, Confocal and STED co-immunofluorescence of CAV3 and TfR1 in ventricular cardiomyocytes. The cartoon of a ventricular cardiomyocyte corresponds with the subcellular image panels on the right showing the confocal and STED data. Dashed boxes indicate high-power magnifications. Scale bars: image panels 1 µm; magnifications 200 nm. **B**, Confocal imaging of living ventricular cardiomyocytes labeled with differic Transferrin-Alexa488. Differic Transferrin signals at the lateral membrane surface of extracellularly labeled ventricular cardiomyocytes are shown. Ventricular cardiomyocytes not exposed to differic Transferrin-Alexa488 were used as negative control. Scale bars: image panels 10 µm. **C**, STED imaging of a living ventricular cardiomyocyte co-labeled with differic Transferrin-Alexa488 and Cholesterol-PEG-KK144 showing differic Transferrin signals adjacent to cholesterol-rich domains at the lateral surface and in transverse tubules. Dashed boxes indicate magnifications. Scale bars: image panels 1 µm; magnifications 200 nm.

## 5.3  CAV3 protein complexes maintain the surface expression of the TfR1

Based on the proposed organization of TfR1 and CAV3 in cholesterol-rich membrane domains, we hypothesized that CAV3 protein complexes are necessary as scaffolds for protein interactions that stabilize the expression of TfR1 in the surface membrane. To test this hypothesis, we generated a human pluripotent stem cell (iPSC) knock-out model, targeting the start codon of CAV3 exon 1 by clustered regularly interspaced short palindromic repeats (CRISPR)/Cas9 (manuscript Supplement Figure 3.13 A). While immunoblotting of WT iPSC-derived cardiomyocyte lysates confirmed robust protein expression each of CAV3 (2912, Abcam) and TfR1 (13-6800, Invitrogen) (Figure 5.3 A), TfR1 expression was significantly decreased in CAV3 knock-out iPSC derived cardiomyocytes (Figure 5.3 A). Furthermore, the expression of TfR1 was analyzed by extracellular surface biotinylation in living iPSC-cardiomyocytes to explore if a global decrease in TfR1 is functionally relevant in the sarcolemma. Biotinylated proteins were enriched by affinity purification and immunoblotting indeed demonstrated a significantly reduced surface expression of TfR1 in human CAV3 knock-out cardiomyocytes (Figure 5.3 B). As expected the negative control, β-Actin (sc-47778, Santa Cruz), was not biotinylated (Figure 5.3 B). Taken together, we showed that CAV3 knock-out disrupts TfR1 expression in the plasma membrane, indicating that CAV3 complexes are required to maintain TfR1 surface expression.

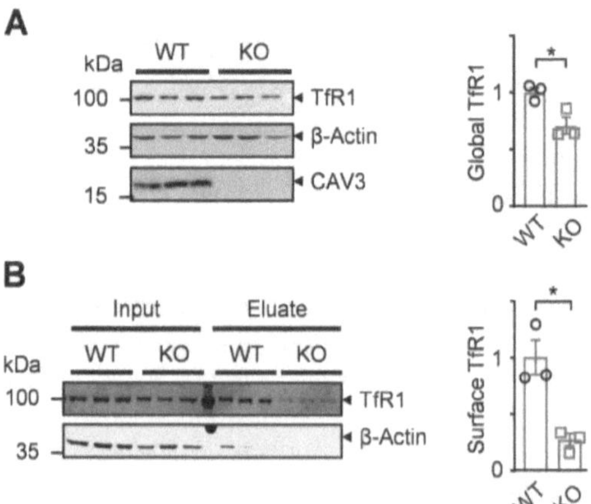


**Figure 5.3 CAV3 knock-out leads to disrupted TfR1 surface expression in human iPSC-cardiomyocytes. A**, Immunoblot analysis of human iPSC-derived CAV3 knock-out cardiomyocytes showed robust expression of CAV3 and TfR1. CAV3 antibody-specific signals were confirmed in CAV3 knock-out iPSC-derived cardiomyocytes. Bar graph showing a significant reduction of global TfR1 expression in CAV3 knock-out cardiomyocytes normalized to β-Actin. n=3. Student's t-test, * p<0.05. **B**, Extracellular protein biotinylation was applied to living human iPSC-derived cardiomyocytes to assess TfR1 surface expression. Biotinylated proteins were enriched by affinity purification and TfR1 was identified by immunoblotting in the eluated fraction. β-Actin immunoblotting was used as negative cytosolic protein labeling control. Bar graph showing a significant loss of surface TfR1 in CAV3 knock-out versus WT cardiomyocytes. n=3. Student's t-test, * p<0.05.

## 5.4 Ratiometric proteomics identifies a loss of McT1 and TfR1 proximity due to the CAV3-S141R mutation

To analyze effects of CAV3 mutations in living neonatal rat cardiomyocytes (NRCMs), we used the same APEX2 labeling strategy established for WT CAV3 (manuscript Figure 3.2 A). Similar to WT CAV3, we achieved stable Isotope Labeling by Amino acids in Cell culture (SILAC) based on incorporation of 96.5% or higher in NRCMs using adenoviral expression (MOI 1) of V5-APEX2-CAV3-S141R (Figure 5.4 A). Furthermore, V5-APEX2-CAV3-S141R expression was confirmed by V5 (R960-25, Invitrogen) immunoblotting (Figure 5.4 B), and V5-APEX2-CAV3-S141R dependent biotinylation of proximal proteins was confirmed by streptavidin (streptavidin-RD680; LI-COR) blotting (Figure 5.4 B). Confocal V5 immunofluorescence imaging confirmed the co-localization of V5-APEX2-CAV3-S141R with native CAV3 (manuscript Supplement Figure 3.14 B). Consistent with a physiological localization pattern in NRCMs, we frequently identified plasma membrane proteins. In comparison to the WT, we identified a small increase of 5 Golgi-associated proteins for CAV3-S141R over 3 Golgi-associated proteins for CAV3-WT (Figure 5.4 C). Importantly, we identified a significant increase of 15 Golgi-associated proteins for CAV3-F97C (manuscript Supplement Figure 3.15 C). Interestingly, similar to the F97C mutation (manuscript Supplement Figure 3.15 C), the number of mitochondrial proteins in the proximity of V5-APEX2-CAV3-S141R were also reduced (Figure 5.4 C). We identified 28 mitochondrial annotated proteins for V5-APEX2-CAV3-S141R, while 46 proteins were identified for WT V5-APEX2-Cav3 (Figure 5.4 C).

As described in the manuscript, blue native polyacrylamidgelelektrophorese (BN-PAGE) analysis showed that the S141R mutation formed high molecular weight complexes similar to the WT (manuscript Figure 3.6 B). Moreover, the S141R mutation preserved the proximity between proteins of the caveolar core complex, while the more peripheral proximity near the core complex with McT1, Ncx1, and TfR1 was disrupted (manuscript Figure 3.7 C). We mapped the interaction networks using the GO terms 'caveolae', 'pyruvate metabolism', and 'iron uptake & transport' based on the STRING database (v11).[142] For the GO term 'caveolae', only the proximity to CAV1 was lost, while the proximity for 'pyruvate metabolism' and 'iron uptake & transport' associated proteins was completely diminished (Figure 5.4 D). These data suggest that V5-APEX2-CAV3-S141R is transported as a core complex to the plasma membrane, however, select peripheral interactions with key metabolic proteins, namely McT1 and TfR1 may become impaired.

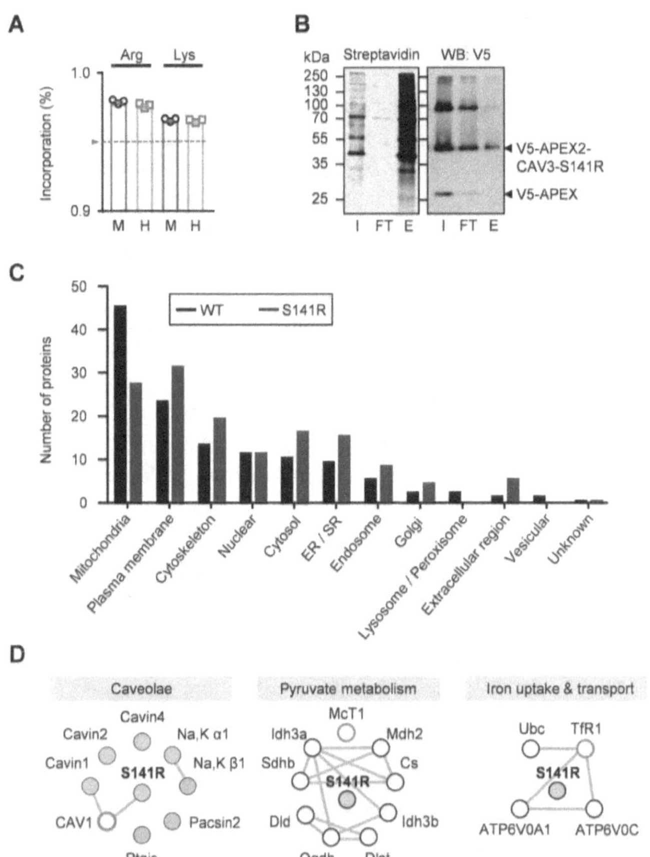





Figure 5.4 **S141R V5-APEX2-CAV3 shows proximity changes with CAV3 interactors. A**, Mass spectrometry (LC-MS/MS) quantified >96 % or higher L-arginine (Arg) and L-lysine (Lys) incorporation. Of note, the y-axis starts at 0.9 to visualize that the SILAC incorporation was >95 % (red line). n=3. **B**, APEX2 biotinylated proteins were captured by streptavidin. I, input; FT, flow through; E, eluate. V5-APEX2-CAV3-S141R and V5-APEX2 expression was confirmed by V5 immunoblotting. n=3. **C**, Bar graph comparing the numerical changes in organelle-specific protein numbers detected for WT and S141R V5-APEX2-CAV3 positive hits. **D**, Analysis of identified proteins for the GO terms 'caveolae', 'pyruvate metabolism', and 'iron uptake & transport' based on the STRING database. Open circles indicate a loss of proximity identification; the red circle highlights the loss of CAV1, McT1 and TfR1 in proximity; blue circles indicate preserved interactions. Grey lines indicate protein interactions with a confidence score of >0.7 or higher.

## 5.5 Differential expression of the CAV1 α- and β-forms in atrial versus ventricular cardiomyocytes

Initially, we identified CAV1 in ventricular cardiomyocytes from adult mouse hearts due to the robust isolation of large cell numbers for the mass spectrometry analysis (manuscript Figure 3.3 B). Interestingly, a recent SWATH-MS analysis identified CAV1 in atrial and ventricular healthy human heart tissues from three adult male individuals.[16] To investigate if CAV1 is expressed in isolated mouse atrial cardiomyocytes, we used immunoblotting and STED nanoscopy. Moreover, to analyze if the CAV1 α- and β-forms, previously identified in skin fibroblasts,[143] are present in cardiomyocytes, we used a CAV1 antibody (Ab17052, Abcam) that detects the epitope shared by both CAV1 α- and β-forms. CAV1 α- and β-forms were robustly detected in atrial cardiomyocytes, while ventricular cardiomyocytes only expressed the α-form (Figure 5.5 A). Even increased signal intensities by 2-fold (Image Studio software, LICOR) did not reveal a band for the β-form in ventricular cardiomyocytes (Figure 5.5 A). Interestingly, we observed two bands for the β-form in atrial cardiomyocytes, suggesting additional posttranslational modifications, which have not been reported for the β-form yet (Figure 5.5 A). The CAV1 antibody specificity was confirmed in CAV1 knock-out mouse hearts (Figure 5.5 A). The α-form was predominantly expressed in atrial cardiomyocytes, while ventricular cardiomyocytes expressed the α-form at more modest (~25 %) levels (Figure 5.5 B). Of note, for a quantitative analysis of the CAV1 expression levels in atrial versus ventricular cardiomyocytes, a further quantitative SWATH-MS analysis is needed. Furthermore, the α-form was identified as the major form in atrial cardiomyocytes, with 2-fold higher expression compared to the β-form (Figure 5.5 B). Importantly, CAV3 expression was not affected in CAV1 knock-out mice (Figure 5.5 C). Taken together, our data indicate that the shorter CAV1 β-form is differentially expressed in atrial cardiomyocytes, suggesting so far unknown CAV1 form specific subcellular or functional roles.

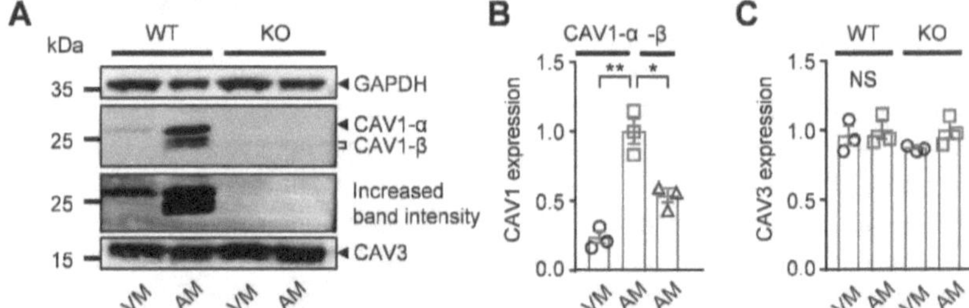



**Figure 5.5 Identification of CAV1 isoforms in atrial and ventricular cardiomyocytes. A**, Immunoblotting detected specific CAV1 signals in atrial (AM) and ventricular cardiomyocytes (VM) of wild-type hearts. CAV1 specificity was confirmed in CAV1 knock-out hearts. Band intensity was increased 2-fold (Image Studio software, LICOR) to confirm the absence of the CAV1 β-form in VM. n=3. **B**, Bar graph comparing the expression levels of CAV1 isoforms in AM and VM. Only AM expressed both CAV1 forms, while the α-form was significantly lower in VM. n=3. One-way-ANOVA, *p<0.05, **p<0.01. **C**, Bar graph showing no significant differences of endogenous CAV3 expression levels between AM and VM in WT or CAV1 knock-out cardiomyocytes, normalized to GAPDH. n=3. One-way-ANOVA.

## 5.6 Distribution of CAV1 versus CAV3 nanodomains in atrial cardiomyocytes

As described in the manuscript (manuscript figure 3.3 D), we identified differential subcellular CAV1 versus CAV3 cluster distributions in the transverse-tubules in ventricular cardiomyocytes. To further investigate the relationship between CAV1 and CAV3 in atrial cardiomyocytes, we used confocal and STED microscopy. Confocal imaging of atrial cardiomyocytes showed CAV1 (Ab17052, Abcam) and CAV3 (Ab2912, Abcam) signals mainly at the lateral membrane and in axial tubules (Figure 5.6 A). In contrast, unspecific cytosolic CAV1 signals not associated with any membrane structures were occasionally apparent in AMs from CAV1 knock-out mice (Figure 5.6 A). STED nanoscopy was used to show the CAV1 and CAV3 cluster signals at superresolution (Figure 5.6 B). CAV1 and CAV3 signals occurred frequently adjacent to each other (Figure 5.6 B). While the presence of CAV1 in cardiomyocytes is still controversially discussed for striated muscles,[18,19,20,21] our data show that CAV1 and CAV3 are both abundantly expressed in atrial and ventricular cardiomyocytes from the same mouse hearts. The differential distribution of CAV1 versus CAV3 clusters indicate independent macromolecular scaffolds that may provide distinct subcellular protein interactions and functions and highlight an atria-specific expression pattern different from the ventricular myocytes.

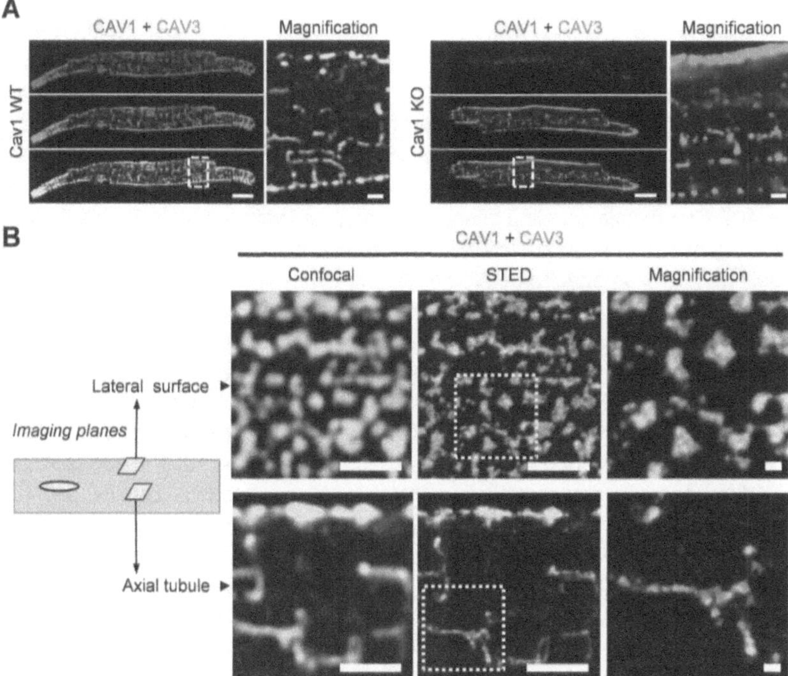



Figure 5.6 **Localization of CAV1 and CAV3 clusters in atrial cardiomyocytes. A,** Confocal imaging of CAV1 and CAV3 clusters in atrial cardiomyocytes showed CAV1 (Ab17052, Abcam) signals at the lateral surface membrane and in axial tubules. The CAV1 signal specificity was confirmed by confocal microscopy of atrial cardiomyocytes isolated from CAV1 knock-out mouse hearts, which also showed some regions with unspecific cytosolic signals. CAV1 signals were detected by using the same CAV1 antibody concentration (1 µg/mL of Ab17052, Abcam). Dashed boxes indicate magnified regions of interest. Scale bars: 10 µm overviews; magnifications 2 µm. **B,** Cartoon of an atrial myocyte corresponding to the subcellular region of interest subjected to confocal and STED imaging. STED nanoscopy of CAV1 and CAV3 clusters in atrial cardiomyocytes resolved frequently adjacent CAV1 and CAV3 clusters at the lateral surface membrane and in axial-tubules. Dashed boxes indicate higher magnifications. Scale bars: confocal microscopy 2 µm; STED nanoscopy 2 µm; STED magnifications 200 nm.

# 6 Discussion

## 6.1 Summary of the results

The aim of this book was to define the cardiac CAV1 and CAV3 protein interactions by live-cell proximity and affinity based mass spectrometry approaches. For live-cell proteomics, we used an engineered Ascorbate PEroXidase (APEX2) as genetic tag for protein biotinylation to initially screen for unknown CAV3 protein interactors in living neonatal rat cardiomyocytes (NRCMs). In addition, to establish a quantitative proximity proteomic technique, the APEX2 proximity labeling assay was combined with a 3-state Stable Isotope Labeling with Amino Acids in Cell Culture (SILAC) workflow. Through APEX2 labeling and mass spectrometry, we identified the NRCM-specific components of the caveolar core complex consistent with earlier studies.[139,140] Importantly, mass spectrometry identified the monocarboxylate transporter (McT1) and the transferrin receptor (TfR1) as new candidates in the nanometric proximity of the CAV3 complex. Surprisingly, we also detected CAV1 in proximity of CAV3, overcoming the historical notion that CAV1 is not expressed in muscle cells.[22] To validate the existence of CAV1 in cardiomyocytes, we used superresolution STED microscopy and quantitative SWATH-MS proteomics, identifying CAV1 as a highly abundant isoform in functional important membrane domains in ventricular cardiomyocytes. Furthermore, immunoblotting revealed a differential expression of distinct CAV1 splice products. While the CAV1 α-form was expressed in both ventricular and atrial cardiomyocytes, the CAV1 β-form was exclusively expressed in atrial cardiomyocytes. Taken together, these data indicate a functional relevance of CAV1 in atrial and ventricular cardiomyocytes. Superresolution microscopy revealed juxta-positioned but not mixed CAV1 and CAV3 cluster signals, indicating an isoform-specific cluster organization in ventricular cardiomyocytes. Moreover, affinity proteomics identified CAV1 versus CAV3 isoform-specific protein interactions. CAV3 specific interactors included the insulin dependent glucose transporter (GluT4), McT1 and TfR1, which are highly relevant for cardiac energy metabolism. Vice versa, aquaporin-1 was identified as specific CAV1 interactor. In line with CAV3-dependent McT1 and TfR1 protein interactions, CAV3 knock-out in human induced pluripotent stem cell (iPSC) derived cardiomyocytes resulted in reduced surface expression of McT1 and TfR1, suggesting a previously unknown role of CAV3 for functional stabilization of these transmembrane proteins in the surface membrane. Furthermore, destabilization of McT1 expression by CAV3 knock-out was associated with reduced extracellular acidification. In conclusion, these data uncovered a previously unknown role of CAV3 for stable surface expression of

McT1 and thus transmembrane proton/lactate shuttling in human iPSC-derived cardiomyocytes.

To explore how potentially pathogenic human CAV3 mutations interfere with CAV3-specific protein interactions, we used proximity based proteomics in NRCMs for V5-APEX2-CAV3-F97C and V5-APEX2-CAV3-S141R. Both the F97C and S141R mutations disrupted the proximity to McT1 and TfR1. However, confocal microscopy in NRCMs revealed that only V5-APEX2-CAV3-F97C resulted in Golgi accumulation and diminished CAV3 oligomerization as shown by Blue Native (BN)-PAGE. In contrast V5-APEX2-CAV3-S141R did not accumulate in the Golgi and formed high molecular weight complexes similar to WT V5-APEX2-CAV3. Furthermore, CAV3-F97C knock-in in iPSC derived cardiomyocytes caused a 97% loss in McT1 surface expression, leading to significantly depressed transmembrane proton export and decreased mitochondrial respiration. Therefore, we propose a novel pathogenic mechanism for the CAV3-F97C mutation leading to impaired metabolite transport, which affects mitochondrial respiration in human cardiomyocytes.

## 6.2 APEX2 proximity assay identified CAV3 protein networks

We used an APEX2[94,95] proximity assay in living NRCMs to identify unknown CAV3 protein interactions based on biotinylation and mass spectrometry. Since protein biotinylation is a rare posttranslational modification in mammalian cells,[96,97] this strategy has been previously used to identify new protein interaction networks in HEK293A and Cos-7 cells.[95,99,144,145] Previously, two protein biotinylation techniques, APEX2[97] and BioID[144], were developed. While APEX2 biotinylates proteins within a radius of approximately 20 nm[97], BioID, based on the biotin protein ligase BirA[144], provides a smaller biotinylation radius of approximately 10 nm.[146] However, the key advantage of APEX2 over BioID is the higher enzymatic activity.[95,145] While APEX2 catalyzes within a 1-min $H_2O_2$ pulse the oxidation of biotin-phenol to a short-lived biotin-phenoxyl radical,[95,145] BirA catalyzes biotin to biotinoyl-5-AMP within several hours (18-24h).[147] The long incubation time increases the chance for unspecific protein detection based on diffusion in the biotinylation radius.[95,145] However, a short treatment with $H_2O_2$, necessary for the APEX2 protocol, can induce oxidative stress, which can lead to changes in gene expression,[148] and to changes in protein interactions.[149] For example, oxidative stress induces the oxidation of redox sensitive cysteines, which may be linked by disulfide bonds to potentially unspecific protein complexes.[149] Despite this potential limitation, the higher enzymatic activity of APEX2 allows for a more effective enrichment of biotinylated proteins.

For APEX2 based biotinylation, we expressed V5-APEX2-CAV3 using an adenoviral transfection strategy. As excessive overexpression of CAV1 in heterologous cell systems was shown to specifically increase the pool of non-caveolar CAV1 in endosomes and interfere with caveolar biogenesis[79,80], we carefully titrated the V5-CAV3-APEX2 expression at levels similar to endogenous CAV3 levels. Confocal imaging of adenoviral transfected NRCMs with a multiplicity of infection of 1 (MOI 1), showed co-localized V5-APEX2-CAV3 with endogenous CAV3 at the plasma membrane, indicating a preserved physiological surface expression of V5-APEX2-CAV3. Moreover co-immunoprecipitation and blue native (BN)-PAGE of V5-CAV3-APEX in NRCMs confirmed that exogenous V5-APEX2-CAV3 formed a high molecular weight hetero-oligomeric complex with endogenous CAV3. In contrast, 3-fold or 10-fold MOI doses led to excessive V5-APEX2-CAV3 overexpression and accumulation in Golgi organelles of NRCMs. Therefore, the titration of V5-APEX2-CAV3 expression down to the lowest effective dose was necessary to overcome the limitations of CAV3 overexpression. However, the use of adenoviral vectors can induce cell stress[150] and the heterogeneity of transfection can influence the V5-APEX2-CAV3 expression, and thus the effective enrichment of biotinylated proteins. Nevertheless, we could show that using this protocol in cardiomyocytes, V5-APEX2-CAV3 maintained essential protein interactions with Cavin1, Pacsin2, Ehd3 and Ehd4, protein components of the multimeric CAV3 core complex. In future studies, genome editing of human stem cells could be used to establish endogenous expression levels of the V5-APEX2-CAV3, to overcome limitations of overexpression systems.

In addition, the APEX2 assay biotinylates proximal proteins prior to any cell lysis, to overcome false negative data due to cell lysis dependent disruption of protein interactions.[95,99] Therefore, APEX2 allows to label high and low affinity and transient interactions, in addition to proximal proteins.[95,99] By that, protein interaction networks can be mapped, in which specific interactions can be further validated and characterized by affinity based interaction approaches. Therefore the strength of APEX2 in combination with affinity proteomics was used to identify CAV3 interactors.

## 6.3 CAV3 stabilizes the surface expression of novel interactors with transmembrane metabolic functions

Combining the APEX2 proximity and affinity proteomics, we identified a novel role of CAV3 as an isoform specific interactor of McT1 and TfR1. STED superresolution microscopy revealed proximal McT1 and TfR1 at CAV3 domains at lateral membranes and transverse (T-)tubules. This is in line with immunogold electron microscopy (EM) studies showing McT1 in rat hearts[151] and TfR1 in mouse hearts[152] in caveolae and T-tubules. Importantly, CAV3 knock-out in iPSC derived cardiomyocytes resulted in a significant decrease of McT1 and TfR1 surface expression, indicating a previous unknown role for functional stabilization of McT1 and TfR1 in the surface membrane. TfR1 facilitates the uptake of monoferric and differic Transferrin via clathrin-mediated endocytosis.[141] Co-labeling of living ventricular cardiomyocytes with diferric Transferrin-Alexa488 and Cholesterol-PEG-KK114 indicated that Transferrin binds to cholesterol-rich membrane domains, which are organized by CAV3 clusters. Therefore, we hypothesized that CAV3 membrane nanodomains serve as a macromolecular scaffold that provides the machinery for cardiomyocyte iron uptake by stable TfR1 surface expression and extracellular receptor presentation. Iron is required for the synthesis of iron co-factors[153], which are essential for oxygen transport[154], DNA synthesis[155] and oxidative phosphorylation.[156] For example, iron-sulfur (Fe-S) clusters are synthesized in mitochondria and facilitate the electron transport chain of mitochondrial inner membrane complexes.[153] Indeed, TfR1 knock-out hearts were previously associated with iron deficiency, leading to enlarged, disrupted mitochondria and reduced enzymatic activity of respiratory complexes I–IV.[156] Moreover, iron deficiency was associated with a reduced production of ATP leading to impaired myocyte contractility and heart failure.[157,158] Therefore, we hypothesized that a reduced TfR1 surface expression induced, by CAV3 knock-out can severely impact mitochondrial respiratory function.

Furthermore, the CAV3 interactor McT1 mediates the proton-coupled transport of small monocarboxylates, particularly lactate and pyruvate, across the plasma membrane.[159] Destabilization of McT1 expression by CAV3 knock-out in iPSC derived cardiomyocytes was associated with reduced extracellular acidification, indicating that CAV3 knock-out impaired the transmembrane proton/lactate shuttling. Lactate import is essential for normal heart function as lactate deprivation was associated with reduced ATP synthesis.[160] During exercise lactate is utilized as a major cardiac energy source, and can account for over 50% of oxygen consumption.[161] Importantly, McT1 protein expression and lactate uptake are significantly upregulated in heart failure.[162] Finally, during ischemia, cardiomyocytes rely significantly on anaerobic glycolysis, which required rapid lactate efflux.[163]

In summary, we proposed a novel function of CAV3 for the organization and stable surface expression of McT1 and TfR1, which are known to be highly relevant for cardiac metabolism.

## 6.4 Metabolic effects of the CAV3-F97C mutation

As discussed in the manuscript, V5-APEX2-CAV3-F97C lost the proximity to McT1 and TfR1. Furthermore, we identified a severe loss of McT1 surface expression in CAV3-F97C knock-in human iPSC-derived cardiomyocytes. As expected for the proton-coupled monocarboxylate metabolite export, which depends on McT1 surface expression[159], the near-complete loss of McT1 surface expression resulted in decreased extracellular acidification, indicating intracellular lactate and proton accumulation. Moreover, a CAV3-F97C knock-in resulted in decreased mitochondrial respiration and less ATP production. ATP is required for muscle relaxation and contraction and for ATP-driven ion pump function, particularly for Na,K-ATPase[164] and SERCA2A[164]. In cardiac ischemia, reduced ATP levels by more than 70 % contribute to increased intracellular concentrations of $Na^+$ and $Ca^{2+}$ ions, which can contribute to cardiac dysfunction and to arrhythmia.[165] Interestingly, pharmacological inhibition of McT1 in tumor cells was associated with decreased glutathione (GSH) synthesis, leading to mitochondrial dysfunction and reduced ATP levels.[166] The mechanism, however, of how increased lactate and proton levels affect GSH synthesis, needs to be further characterized. Of note, as discussed above, the impact of CAV3-F97C knock-in on the mitochondrial respiratory function could additionally originate from iron deficiency, induced by reduced TfR1 surface expression, as shown by CAV3 knock-out. The impact of CAV3-F97C knock-in on TfR1 needs to be further investigated by future studies.

## 6.5 Pathological effects of the human CAV3-F97C and -S141R mutations

We showed that V5-APEX2-CAV3-F97C protein accumulated in the Golgi and disrupted the biogenesis of trafficking-competent oligomeric complexes. Therefore, we propose that the C V5-APEX2-CAV3-F97 Golgi accumulation caused a loss of TfR1 and McT1 in the surface membrane of NRCMs. Similarly, V5-APEX2-CAV3-S141R proximity biotinylation showed a loss of TfR1 and Mct1 proximity. However, based on confocal imaging in NRCMs, V5-APEX2-CAV3-S141R did not accumulate in the Golgi and was correctly transported to the plasma membrane. Moreover, BN-PAGE analysis after overexpression of V5-APEX2-CAV3-S141R in HEK293A cells revealed, that V5-APEX2-CAV3-S141R formed high molecular weight complexes similar to WT V5-APEX2-CAV3. In line with these results, V5-APEX2-CAV3-S141R did not affect the proximity of typical proteins of the caveolar core complex or of caveolae associated proteins. Moreover, we identified multiple plasma membrane annotated proteins in proximity to V5-APEX2-CAV3-S141R. Interestingly, immunofluorescence of muscle tissue biopsies from hyperCKemia patients with the CAV3-T78M mutation also showed that the CAV3-T78M protein is correctly transported to the plasma membrane.[167] However, the pathomechanism of CAV3-T78M mutation remains unclear.[167] Taken together, our proximity proteomic data suggest that the CAV3-S141R mutation may have an impact on the cell physiology in NRCMs through impaired protein interactions. In contrast to CAV3-F97C mutation, S141R did not impair the biogenesis of trafficking competent oligomers to the plasma membrane. However, the precise role of the CAV3-S141R mutation on transmembrane proteins needs to be further characterized by future studies.

## 6.6 Function of CAV3 in GluT4 mediated glucose uptake

In addition to the metabolite transporters McT1 and TfR1, affinity proteomics identified the insulin dependent glucose transporter (GluT4) as CAV3 protein interactor in mouse ventricular cardiomyocytes. The GluT4-CAV3 interaction was previously shown by immunoprecipitation in isoflurane treated adult rat cardiomyocytes.[168] GluT4 is an insulin dependent glucose transporter, which is expressed in adipocytes, skeletal and cardiac muscles.[169] In the heart, GluT4 knock-out diminished the insulin mediated glucose uptake, resulting in cardiac hypertrophy.[170] However, the influence of CAV3 on GluT4 mediated glucose uptake in cardiomyocytes is not well understood. The functional relevance of the GluT4-CAV3 interaction has been previously shown in skeletal myocytes, in which CAV3 facilitates the insulin dependent transport of GluT4 to the surface membrane.[171] Confocal microscopy in skeletal muscle fibers revealed GluT4 in proximity to CAV3 domains at the lateral surface and in T-tubules.[172] Furthermore, CAV3 knock-out resulted in decreased insulin-stimulated glucose uptake in skeletal muscle of CAV3 knock-out mice.[173] Strikingly, the CAV3 mutation P104L which was associated with limb girdle muscular dystrophy, resulted in a decreased glucose uptake in cultured skeletal myotubes through diminished insulin-induced surface expression of GluT4.[122] Consequently, we hypothesized a similar functional role of CAV3 for the regulation of glucose uptake in cardiomyocytes. Further analysis of the CAV3-GluT4 interaction in the CAV3 knock-out and F97C knock-in stem cell models can be used in future studies to validate the proposed functional role in human cardiomyocytes.

## 6.7 Differential expression of CAV1 splice forms in cardiomyocytes

We identified CAV1 by quantitative SWATH-MS as highly abundant isoform in ventricular adult mouse cardiomyocytes, overcoming the historical notion that CAV1 is only expressed in non-muscle cells.[17] Accordingly, a recent SWATH-MS analysis identified CAV1 in atrial and ventricular healthy human heart tissues.[16] Interestingly, two CAV1 splice forms were previously identified in skin fibroblasts, a full length α-form and a 31 amino acid N-terminally truncated β-form.[143] As the CAV1 β-form is identical with the CAV1 α-form, except for the N-terminal extension, both splice forms cannot be distinguished by mass spectrometry. Therefore, we used immunoblotting to analyze the presence of CAV1 α- and β-forms in atrial and ventricular cardiomyocytes. By using a CAV1 antibody detecting both CAV1 α- and β-forms, we identified the CAV1 α- and β-form in atrial cardiomyocytes, while only the CAV1 α-form was expressed in ventricular cardiomyocytes. A differential expression of CAV1 α- and β-forms was previously shown by comparing lysates of endothelial with alveolar epithelial cells of rat lungs.[174] While lung endothelial cells expressed only the α-form, lung alveolar epithelial cells expressed the CAV1 β- form.[174] Therefore, the differentially expression of CAV1 splice forms were proposed to have potentially unique physiological functions.[143]

As the β-form lacks the first 30 amino acids, which are required for the Src mediated phosphorylation at tyrosin 14 of the alpha-form,[175] functional differences for the CAV1 α- and β-forms were proposed.[176] Phosphorylation of CAV1 at tyrosin 14 promotes caveolae-mediated endocytosis[177] and the recruitment of proteins containing the Src Homology 2 (SH2)-domain, which modulate the receptor tyrosine kinase pathway.[178,179] Therefore the CAV1 α-form was proposed to regulate receptor tyrosine kinases,[178,179] while a specific function for the CAV1 β-form remains unknown.[143] Future studies may explore the precise function of the CAV1 forms in cardiomyocytes.

## 6.8 Substantial role of CAV1 for cardiomyocyte function

As the presence of CAV1 in cardiomyocytes remained controversial,[18,19,20,21] the adverse effects of CAV1 knock-out on the mouse heart where solely related to cardiac fibroblasts and endothelial cells.[100,101] Consequently, the effects of human CAV1 mutations were never considered in cardiomyocytes. Recently, a CAV1 deficiency was proposed as cause of atrial fibrillation.[114,112,113] Thus, our confirmation of CAV1 expression in atrial cardiomyocytes will open new perspectives about atrial myocytes in health and disease. Atrial tissue samples from patients with atrial fibrillation also showed a reduced CAV1 expression, while the expression of CAV2 and CAV3 was not changed.[114] Furthermore, genome-wide studies identified one intronic single nucleotide polymorphism (SNP) in the CAV1 gene (rs3807989), which was proposed to contribute to the risk of atrial fibrillation.[112,113] Given that the human CAV1-P132L mutation associated with breast cancer was demonstrated to disrupt caveolae biogenesis and to accumulate in the Golgi[79], similar to the human CAV3-F97C mutation, we anticipate that CAV1 has an important role in cardiomyocyte function.

## 6.9 Potential role of CAV1 and aquaporin-1 interaction

STED superresolution imaging in ventricular cardiomyocytes revealed that CAV1 and CAV3 signals are not co-localized. The subcellular CAV1 and CAV3 clusters frequently occurred adjacent, yet separate to each other. Therefore, we propose CAV1 and CAV3 isoform-specific macromolecular complexes. Supportingly, SWATH affinity proteomics and reciprocal co-immunoprecipitation experiments identified CAV1 and CAV3 isoform specific protein interactions. Aquaporin-1 was identified as exclusive CAV1 interactor in ventricular cardiomyocytes by SWATH affinity proteomics and reciprocal co-immunoprecipitation. Previous studies identified aquaporin-1 as a bidirectional water channel, which facilitates the selective transport of $H_2O$ molecules across cell membranes according to the prevailing osmotic gradient.[180] Consistent with our affinity proteomics analysis, previous immunogold EM studies demonstrated that aquaporin-1 is localized in CAV1 containing caveolae in endothelial cells.[181] This may indicate that CAV1 has a stabilizing role for aquaporin-1 surface expression in the plasma membrane. Accordingly, aquaporin-1 expression was reduced in rat lung tissue lysates after CAV1 protein knock-down of 87% by small interfering RNA (siRNA).[182] Additionally, aquaporin-1 was identified in human, rat and mouse heart tissue lysates by immunoblotting and aquaporin-1 knock-out was associated with significant reduction in water permeability in membrane vesicles of mouse hearts and impaired osmotic homeostasis.[183] An impaired osmotic homeostasis can indirectly affect cellular ionic concentrations and thus cardiac electrophysiology.[184] Interestingly, increased intracellular lactate levels due to ischemia created a strong inwardly directed osmotic gradient,[185,186] which led to swelling of cardiomyocytes and cell damage.[186] Together with our data showing a CAV3-dependent stabilization of McT1-dependent lactate transport and CAV1 interactions with aquaporin-1, we hypothesize that both CAV1 and CAV3 have important roles for cardiac stress adaptation.

## 6.10 Summary and Outlook

The original aim of this book was to identify previously unknown cardiac CAV3 protein interactions preferentially through unbiased proteomic methods. Combining the strengths of proximity and affinity proteomics, we established CAV1- versus CAV3-specific interactors. In this context, we showed that exogenous V5-APEX2-CAV3 expression at levels similar to endogenous CAV3 levels is an important prerequisite to detect protein interactions with the multimeric CAV3 core complexes. In contrast, excessive V5-APEX2-CAV3 overexpression in NRCMs disrupted the physiological surface expression of the CAV3 core complex likely due to Golgi accumulation and lack of trafficking-competent protein-lipid complexes. For CAV3 we identified McT1 and TfR1 as new isoform-specific protein interactors in ventricular cardiomyocytes, which are relevant for the cardiac energy metabolism. CAV3 knock-out in human stem cells revealed, that CAV3 is required for surface expression of both McT1 and TfR1. Importantly, F97C knock-in in human stem cells destabilized the surface expression of McT1 and thus the lactate-coupled proton export, which led to reduced extracellular acidification, mitochondrial respiration and ATP production.

Moreover, in contrast to V5-APEX2-CAV3-S141R, V5-APEX2-CAV3-F97C expression in NRCMs resulted in a loss of physiological proximity with Cavin1, Pacsin2, Ehd3 and Ehd4 protein components of the multimeric CAV3 core complex. Interestingly, proximity proteomics also showed that both V5-APEX2-CAV3-F97C and V5-APEX2-CAV3-S141R resulted in a loss of proximity with McT1 and TfR1, indicating that both CAV3 variants impaired the same isoform-specific protein interactions.

Furthermore, quantitative proteomics and superresolution imaging showed that CAV1 clusters exist in mouse ventricular cardiomyocytes, which are juxta-positioned to CAV3 clusters in T-tubules. In line with our hypothesis that CAV1 and CAV3 provide differential protein interactions, we identified aquaporin-1 as a specific CAV1 interactor. This may indicate a novel role also for CAV1 to stabilize the surface expression of aquaporin-1. In addition, we showed that the CAV1 α- and β-forms were differentially expressed in atrial and ventricular cardiomyocytes, indicating so far unknown specific functional roles.

In summary, the combination of proximity and affinity proteomics, revealed previously unknown cardiac CAV1 and CAV3 protein interactions, providing a strategy for systematic functional analysis. The identified cardiac CAV1 and CAV3 protein interactions can be used in future studies to explore the molecular impact of human CAV mutations in the context of cardiac muscle function, as the identified CAV3 interactions McT1 and TfR1 provide an important role in energy metabolism. Taken together, the validation of CAV1 and CAV3 protein networks represent an important direction to identify and define isoform-specific protein interactions in cardiac cell biology.

# 7 References


1. Ueda K. Contractile function of the heart. Cardiac function in relation to cardiac anatomy. *Nippon Rinsho.* 1975;33:2223–2229.

2. Plačkić J, Kockskämper J. Isolation of Atrial and Ventricular Cardiomyocytes for In Vitro Studies. *Methods Mol Biol.* 2018;1816:39–54.

3. Brandenburg S, Arakel EC, Schwappach B, Lehnart SE. The molecular and functional identities of atrial cardiomyocytes in health and disease. *Biochim Biophys Acta.* 2016;1863:1882–1893.

4. O'Connell TD, Rodrigo MC, Simpson PC. Isolation and culture of adult mouse cardiac myocytes. *Methods Mol Biol.* 2007;357:271–296.

5. Parton RG, Way M, Zorzi N, Stang E. Caveolin-3 Associates with Developing T-tubules during Muscle Differentiation. *J Cell Biol.* 1997;136:137–154.

6. Rougier J-S, Abriel H. Cardiac voltage-gated calcium channel macromolecular complexes. *Biochimica et Biophysica Acta (BBA) - Molecular Cell Research.* 2016;1863:1806–1812.

7. Eisner DA, Caldwell JL, Kistamás K, Trafford AW. Calcium and Excitation-Contraction Coupling in the Heart. *Circ Res.* 2017;121:181–195.

8. Brandenburg S, Pawlowitz J, Fakuade FE, Kownatzki-Danger D, Kohl T, Mitronova GY, Scardigli M, Neef J, Schmidt C, Wiedmann F, Pavone FS, Sacconi L, Kutschka I, Sossalla S, Moser T, Voigt N, Lehnart SE. Axial Tubule Junctions Activate Atrial Ca2+ Release Across Species. *Front Physiol.* 2018;9.

9. McNutt NS, Fawcett DW. The ultrastructure of the cat myocardium. II. Atrial muscle. *J Cell Biol.* 1969;42:46–67.

10. Bootman MD, Smyrnias I, Thul R, Coombes S, Roderick HL. Atrial cardiomyocyte calcium signalling. *Biochimica et Biophysica Acta (BBA) - Molecular Cell Research.* 2011;1813:922–934.

11. Brandenburg S, Kohl T, Williams GSB, Gusev K, Wagner E, Rog-Zielinska EA, Hebisch E, Důra M, Didié M, Gotthardt M, Nikolaev VO, Hasenfuß G, Kohl P, Ward CW, Lederer WJ, Lehnart SE. Axial tubule junctions control rapid calcium signaling in atria. *The Journal of clinical investigation.* 2016;126:3999–4015.

12. Barth AS, Merk S, Arnoldi E, Zwermann L, Kloos P, Gebauer M, Steinmeyer K, Bleich M, Kääb S, Pfeufer A, Uberfuhr P, Dugas M,

Steinbeck G, Nabauer M. Functional profiling of human atrial and ventricular gene expression. *Pflugers Arch.* 2005;450:201–208.

13.  Bao ZZ, Bruneau BG, Seidman JG, Seidman CE, Cepko CL. Regulation of chamber-specific gene expression in the developing heart by Irx4. *Science.* 1999;283:1161–1164.

14.  Koibuchi N, Chin MT. CHF1/Hey2 plays a pivotal role in left ventricular maturation through suppression of ectopic atrial gene expression. *Circ Res.* 2007;100:850–855.

15.  Wu S, Cheng C-M, Lanz RB, Wang T, Respress JL, Ather S, Chen W, Tsai S-J, Wehrens XHT, Tsai M-J, Tsai SY. Atrial Identity Is Determined by a COUP-TFII Regulatory Network. *Developmental Cell.* 2013;25:417–426.

16.  Doll S, Dreßen M, Geyer PE, Itzhak DN, Braun C, Doppler SA, Meier F, Deutsch M-A, Lahm H, Lange R, Krane M, Mann M. Region and cell-type resolved quantitative proteomic map of the human heart. *Nat Commun.* 2017;8:1–13.

17.  Liu P, Rudick M, Anderson RGW. Multiple Functions of Caveolin-1. *J Biol Chem.* 2002;277:41295–41298.

18.  Robenek H, Weissen-Plenz G, Severs NJ. Freeze-fracture replica immunolabelling reveals caveolin-1 in the human cardiomyocyte plasma membrane. *Journal of Cellular and Molecular Medicine.* 2008;12:2519–2521.

19.  Head BP, Patel HH, Roth DM, Murray F, Swaney JS, Niesman IR, Farquhar MG, Insel PA. Microtubules and Actin Microfilaments Regulate Lipid Raft/Caveolae Localization of Adenylyl Cyclase Signaling Components. *J Biol Chem.* 2006;281:26391–26399.

20.  Patel HH, Tsutsumi YM, Head BP, Niesman IR, Jennings M, Horikawa Y, Huang D, Moreno AL, Patel PM, Insel PA, Roth DM. Mechanisms of cardiac protection from ischemia/reperfusion injury: a role for caveolae and caveolin-1. *The FASEB Journal.* 2007;21:1565–1574.

21.  Volonte D, McTiernan CF, Drab M, Kasper M, Galbiati F. Caveolin-1 and caveolin-3 form heterooligomeric complexes in atrial cardiac myocytes that are required for doxorubicin-induced apoptosis. *American Journal of Physiology-Heart and Circulatory Physiology.* 2008;294:H392–H401.

22.  Parton RG. Caveolae: Structure, Function, and Relationship to Disease. *Annu Rev Cell Dev Biol.* 2018;34:111–136.

23. Parton RG, Hanzal-Bayer M, Hancock JF. Biogenesis of caveolae: a structural model for caveolin-induced domain formation. *Journal of Cell Science*. 2006;119:787–796.

24. Hayer A, Stoeber M, Bissig C, Helenius A. Biogenesis of caveolae: stepwise assembly of large caveolin and cavin complexes. *Traffic*. 2010;11:361–382.

25. Pol A, Martin S, Fernández MA, Ingelmo-Torres M, Ferguson C, Enrich C, Parton RG. Cholesterol and fatty acids regulate dynamic caveolin trafficking through the Golgi complex and between the cell surface and lipid bodies. *Mol Biol Cell*. 2005;16:2091–2105.

26. González Coraspe JA, Weis J, Anderson ME, Münchberg U, Lorenz K, Buchkremer S, Carr S, Zahedi RP, Brauers E, Michels H, Sunada Y, Lochmüller H, Campbell KP, Freier E, Hathazi D, Roos A. Biochemical and pathological changes result from mutated Caveolin-3 in muscle. *Skeletal Muscle*. 2018;8:28.

27. Lee H, Park DS, Razani B, Russell RG, Pestell RG, Lisanti MP. Caveolin-1 Mutations (P132L and Null) and the Pathogenesis of Breast Cancer: Caveolin-1 (P132L) Behaves in a Dominant-Negative Manner and Caveolin-1 (−/−) Null Mice Show Mammary Epithelial Cell Hyperplasia. *The American Journal of Pathology*. 2002;161:1357–1369.

28. Fra AM, Williamson E, Simons K, Parton RG. De novo formation of caveolae in lymphocytes by expression of VIP21-caveolin. *Proc Natl Acad Sci USA*. 1995;92:8655–8659.

29. Echarri A, Pozo MAD. Caveolae – mechanosensitive membrane invaginations linked to actin filaments. *J Cell Sci*. 2015;128:2747–2758.

30. Sinha B, Köster D, Ruez R, Gonnord P, Bastiani M, Abankwa D, Stan RV, Butler-Browne G, Vedie B, Johannes L, Morone N, Parton RG, Raposo G, Sens P, Lamaze C, Nassoy P. Cells Respond to Mechanical Stress by Rapid Disassembly of Caveolae. *Cell*. 2011;144:402–413.

31. Pfeiffer ER, Wright AT, Edwards AG, Stowe JC, McNall K, Tan J, Niesman I, Patel HH, Roth DM, Omens JH, McCulloch AD. Caveolae in ventricular myocytes are required for stretch-dependent conduction slowing. *J Mol Cell Cardiol*. 2014;76:265–274.

32. Kordylewski L, Goings GE, Page E. Rat atrial myocyte plasmalemmal caveolae in situ. Reversible experimental increases in caveolar size and in surface density of caveolar necks. *Circ Res*. 1993;73:135–146.

33. Hagiwara Y, Sasaoka T, Araishi K, Imamura M, Yorifuji H, Nonaka I, Ozawa E, Kikuchi T. Caveolin-3 deficiency causes muscle degeneration in mice. *Hum Mol Genet.* 2000;9:3047–3054.

34. Pelkmans L, Zerial M. Kinase-regulated quantal assemblies and kiss-and-run recycling of caveolae. *Nature.* 2005;436:128–133.

35. Park D, Woodman S, Schubert W, Cohen A, Frank P, Chandra M, Shirani J, Razani B, Tang B, Jelicks L, Factor S, Weiss L, Tanowitz H, Lisanti M. Caveolin-1/3 Double-Knockout Mice Are Viable, but Lack Both Muscle and Non-Muscle Caveolae, and Develop a Severe Cardiomyopathic Phenotype. *The American journal of pathology.* 2002;160:2207–17.

36. Mundy DI, Machleidt T, Ying Y, Anderson RGW, Bloom GS. Dual control of caveolar membrane traffic by microtubules and the actin cytoskeleton. *J Cell Sci.* 2002;115:4327–4339.

37. Parton R, Joggerst B, Simons K. Regulated internalization of caveolae. *The Journal of cell biology.* 1995;127:1199–215.

38. Hertzog M, Monteiro P, Dez GL, Chavrier P. Exo70 Subunit of the Exocyst Complex Is Involved in Adhesion-Dependent Trafficking of Caveolin-1. *PLOS ONE.* 2012;7:e52627.

39. Rothberg KG, Heuser JE, Donzell WC, Ying Y-S, Glenney JR, Anderson RGW. Caveolin, a protein component of caveolae membrane coats. *Cell.* 1992;68:673–682.

40. Echarri A, Muriel O, Pavón DM, Azegrouz H, Escolar F, Terrón MC, Sanchez-Cabo F, Martínez F, Montoya MC, Llorca O, Del Pozo MA. Caveolar domain organization and trafficking is regulated by Abl kinases and mDia1. *J Cell Sci.* 2012;125:3097–3113.

41. Lisanti MP, Scherer PE, Tang Z, Sargiacomo M. Caveolae, caveolin and caveolin-rich membrane domains: a signalling hypothesis. *Trends Cell Biol.* 1994;4:231–235.

42. Kogo H, Fujimoto T. Caveolin-1 isoforms are encoded by distinct mRNAs. Identification Of mouse caveolin-1 mRNA variants caused by alternative transcription initiation and splicing. *FEBS Lett.* 2000;465:119–123.

43. Ramirez MI, Pollack L, Millien G, Cao YX, Hinds A, Williams MC. The α-Isoform of Caveolin-1 Is a Marker of Vasculogenesis in Early Lung Development. *J Histochem Cytochem.* 2002;50:33–42.

44. Glenney JR, Soppet D. Sequence and expression of caveolin, a protein component of caveolae plasma membrane domains phosphorylated on

tyrosine in Rous sarcoma virus-transformed fibroblasts. *Proc Natl Acad Sci USA*. 1992;89:10517–10521.

45.  Schlegel A, Schwab RB, Scherer PE, Lisanti MP. A Role for the Caveolin Scaffolding Domain in Mediating the Membrane Attachment of Caveolin-1: THE CAVEOLIN SCAFFOLDING DOMAIN IS BOTH NECESSARY AND SUFFICIENT FOR MEMBRANE BINDING *IN VITRO*. *J Biol Chem*. 1999;274:22660–22667.

46.  Galbiati F, Engelman JA, Volonte D, Zhang XL, Minetti C, Li M, Hou H, Kneitz B, Edelmann W, Lisanti MP. Caveolin-3 null mice show a loss of caveolae, changes in the microdomain distribution of the dystrophin-glycoprotein complex, and t-tubule abnormalities. *J Biol Chem*. 2001;276:21425–21433.

47.  Cao H, Courchesne WE, Mastick CC. A Phosphotyrosine-dependent Protein Interaction Screen Reveals a Role for Phosphorylation of Caveolin-1 on Tyrosine 14 RECRUITMENT OF C-TERMINAL Src KINASE. *J Biol Chem*. 2002;277:8771–8774.

48.  Whiteley G, Collins RF, Kitmitto A. Characterization of the Molecular Architecture of Human Caveolin-3 and Interaction with the Skeletal Muscle Ryanodine Receptor. *J Biol Chem*. 2012;287:40302–40316.

49.  Dietzen DJ, Hastings WR, Lublin DM. Caveolin is palmitoylated on multiple cysteine residues. Palmitoylation is not necessary for localization of caveolin to caveolae. *J Biol Chem*. 1995;270:6838–6842.

50.  Parat M-O, Stachowicz RZ, Fox PL. Oxidative stress inhibits caveolin-1 palmitoylation and trafficking in endothelial cells. *Biochem J*. 2002;361:681–688.

51.  Fujimoto T, Kogo H, Nomura R, Une T. Isoforms of caveolin-1 and caveolar structure. *Journal of Cell Science*. 2000;113:3509–3517.

52.  Williams TM, Lisanti MP. The Caveolin genes: from cell biology to medicine. *Annals of Medicine*. 2004;36:584–595.

53.  Das K, Lewis RY, Scherer PE, Lisanti MP. The Membrane-spanning Domains of Caveolins-1 and -2 Mediate the Formation of Caveolin Hetero-oligomers IMPLICATIONS FOR THE ASSEMBLY OF CAVEOLAE MEMBRANES IN VIVO. *J Biol Chem*. 1999;274:18721–18728.

54.  Parolini I, Sargiacomo M, Galbiati F, Rizzo G, Grignani F, Engelman JA, Okamoto T, Ikezu T, Scherer PE, Mora R, Rodriguez-Boulan E, Peschle C, Lisanti MP. Expression of Caveolin-1 Is Required for the Transport of Caveolin-2 to the Plasma Membrane: RETENTION OF CAVEOLIN-2 AT

THE LEVEL OF THE GOLGI COMPLEX. *J Biol Chem*. 1999;274:25718–25725.

55. Kovtun O, Tillu VA, Ariotti N, Parton RG, Collins BM. Cavin family proteins and the assembly of caveolae. *J Cell Sci*. 2015;128:1269–1278.

56. Gambin Y, Ariotti N, McMahon K-A, Bastiani M, Sierecki E, Kovtun O, Polinkovsky ME, Magenau A, Jung W, Okano S, Zhou Y, Leneva N, Mureev S, Johnston W, Gaus K, Hancock JF, Collins BM, Alexandrov K, Parton RG. Single-molecule analysis reveals self assembly and nanoscale segregation of two distinct cavin subcomplexes on caveolae. *Elife*. 2013;3:e01434.

57. Ludwig A, Howard G, Mendoza-Topaz C, Deerinck T, Mackey M, Sandin S, Ellisman MH, Nichols BJ. Molecular Composition and Ultrastructure of the Caveolar Coat Complex. *PLoS Biol*. 2013;11:e1001640.

58. Parton RG, Tillu VA, Collins BM. Caveolae. *Current Biology*. 2018;28:R402–R405.

59. Hill MM, Bastiani M, Luetterforst R, Kirkham M, Kirkham A, Nixon SJ, Walser P, Abankwa D, Oorschot VMJ, Martin S, Hancock JF, Parton RG. PTRF-Cavin, a conserved cytoplasmic protein required for caveola formation and function. *Cell*. 2008;132:113–124.

60. Daumke O, Lundmark R, Vallis Y, Martens S, Butler PJG, McMahon HT. Architectural and mechanistic insights into an EHD ATPase involved in membrane remodelling. *Nature*. 2007;449:923–927.

61. Morén B, Shah C, Howes MT, Schieber NL, McMahon HT, Parton RG, Daumke O, Lundmark R. EHD2 regulates caveolar dynamics via ATP-driven targeting and oligomerization. *Mol Biol Cell*. 2012;23:1316–1329.

62. Hansen CG, Howard G, Nichols BJ. Pacsin 2 is recruited to caveolae and functions in caveolar biogenesis. *Journal of Cell Science*. 2011;124:2777–2785.

63. Seemann E, Sun M, Krueger S, Tröger J, Hou W, Haag N, Schüler S, Westermann M, Huebner CA, Romeike B, Kessels MM, Qualmann B. Deciphering caveolar functions by syndapin III KO-mediated impairment of caveolar invagination. *eLife*. 2017;6:e29854.

64. Oh P, McIntosh DP, Schnitzer JE. Dynamin at the neck of caveolae mediates their budding to form transport vesicles by GTP-driven fission from the plasma membrane of endothelium. *J Cell Biol*. 1998;141:101–114.

65. Kwon H, Lee J, Jeong K, Jang D, Pak Y. Fatty acylated caveolin-2 is a substrate of insulin receptor tyrosine kinase for insulin receptor substrate-1-directed signaling activation. *Biochimica et Biophysica Acta (BBA) - Molecular Cell Research*. 2015;1853:1022–1034.

66. Vinten J, Johnsen AH, Roepstorff P, Harpøth J, Tranum-Jensen J. Identification of a major protein on the cytosolic face of caveolae. *Biochim Biophys Acta*. 2005;1717:34–40.

67. Lamaze C, Tardif N, Dewulf M, Vassilopoulos S, Blouin CM. The caveolae dress code: structure and signaling. *Curr Opin Cell Biol*. 2017;47:117–125.

68. Hansen CG, Shvets E, Howard G, Riento K, Nichols BJ. Deletion of cavin genes reveals tissue-specific mechanisms for morphogenesis of endothelial caveolae. *Nat Commun*. 2013;4:1–13.

69. Nabi IR. Cavin fever: regulating caveolae. *Nat Cell Biol*. 2009;11:789–791.

70. Ogata T, Ueyama T, Isodono K, Tagawa M, Takehara N, Kawashima T, Harada K, Takahashi T, Shioi T, Matsubara H, Oh H. MURC, a muscle-restricted coiled-coil protein that modulates the Rho/ROCK pathway, induces cardiac dysfunction and conduction disturbance. *Molecular and cellular biology*. 2008;28:3424–3436.

71. Ogata T, Naito D, Nakanishi N, Hayashi YK, Taniguchi T, Miyagawa K, Hamaoka T, Maruyama N, Matoba S, Ikeda K, Yamada H, Oh H, Ueyama T. MURC/Cavin-4 facilitates recruitment of ERK to caveolae and concentric cardiac hypertrophy induced by $\alpha$1-adrenergic receptors. *Proc Natl Acad Sci USA*. 2014;111:3811–3816.

72. Bastiani M, Liu L, Hill MM, Jedrychowski MP, Nixon SJ, Lo HP, Abankwa D, Luetterforst R, Fernandez-Rojo M, Breen MR, Gygi SP, Vinten J, Walser PJ, North KN, Hancock JF, Pilch PF, Parton RG. MURC/Cavin-4 and cavin family members form tissue-specific caveolar complexes. *J Cell Biol*. 2009;185:1259–1273.

73. Yeow I, Howard G, Chadwick J, Mendoza-Topaz C, Hansen CG, Nichols BJ, Shvets E. EHD Proteins Cooperate to Generate Caveolar Clusters and to Maintain Caveolae during Repeated Mechanical Stress. *Current Biology*. 2017;27:2951.

74. Senju Y, Rosenbaum E, Shah C, Hamada-Nakahara S, Itoh Y, Yamamoto K, Hanawa-Suetsugu K, Daumke O, Suetsugu S. Phosphorylation of PACSIN2 by protein kinase C triggers the removal of caveolae from the plasma membrane. *J Cell Sci*. 2015;128:2766–2780.

75. Vatta M, Ackerman MJ, Ye B, Makielski JC, Ughanze EE, Taylor EW, Tester DJ, Balijepalli RC, Foell JD, Li Z, Kamp TJ, Towbin JA. Mutant Caveolin-3 Induces Persistent Late Sodium Current and Is Associated With Long-QT Syndrome. *Circulation*. 2006;114:2104–2112.

76. Gottlieb-Abraham E, Shvartsman DE, Donaldson JC, Ehrlich M, Gutman O, Martin GS, Henis YI. Src-mediated caveolin-1 phosphorylation affects the targeting of active Src to specific membrane sites. *Mol Biol Cell*. 2013;24:3881–3895.

77. Balijepalli RC, Foell JD, Hall DD, Hell JW, Kamp TJ. Localization of cardiac L-type Ca2+ channels to a caveolar macromolecular signaling complex is required for β2-adrenergic regulation. *Proc Natl Acad Sci U S A*. 2006;103:7500–7505.

78. Gazzerro E, Sotgia F, Bruno C, Lisanti MP, Minetti C. Caveolinopathies: from the biology of caveolin-3 to human diseases. *Eur J Hum Genet*. 2010;18:137–145.

79. Han B, Copeland CA, Tiwari A, Kenworthy AK. Assembly and Turnover of Caveolae: What Do We Really Know? *Front Cell Dev Biol*. 2016;4.

80. Parton RG, del Pozo MA. Caveolae as plasma membrane sensors, protectors and organizers. *Nat Rev Mol Cell Biol*. 2013;14:98–112.

81. Nystrom FH, Chen H, Cong LN, Li Y, Quon MJ. Caveolin-1 interacts with the insulin receptor and can differentially modulate insulin signaling in transfected Cos-7 cells and rat adipose cells. *Mol Endocrinol*. 1999;13:2013–2024.

82. Yamamoto M, Toya Y, Schwencke C, Lisanti MP, Myers MG, Ishikawa Y. Caveolin Is an Activator of Insulin Receptor Signaling. *J Biol Chem*. 1998;273:26962–26968.

83. Feron O, Belhassen L, Kobzik L, Smith TW, Kelly RA, Michel T. Endothelial nitric oxide synthase targeting to caveolae. Specific interactions with caveolin isoforms in cardiac myocytes and endothelial cells. *J Biol Chem*. 1996;271:22810–22814.

84. Feron O, Dessy C, Opel DJ, Arstall MA, Kelly RA, Michel T. Modulation of the endothelial nitric-oxide synthase-caveolin interaction in cardiac myocytes. Implications for the autonomic regulation of heart rate. *J Biol Chem*. 1998;273:30249–30254.

85. Langlois S, Cowan KN, Shao Q, Cowan BJ, Laird DW. Caveolin-1 and -2 Interact with Connexin43 and Regulate Gap Junctional Intercellular Communication in Keratinocytes. *MBoC*. 2008;19:912–928.

86.   Liu L, Li Y, Lin J, Liang Q, Sheng X, Wu J, Huang R, Liu S, Li Y. Connexin43 interacts with Caveolin-3 in the heart. *Mol Biol Rep.* 2010;37:1685–1691.

87.   Cai T, Wang H, Chen Y, Liu L, Gunning WT, Quintas LEM, Xie Z-J. Regulation of caveolin-1 membrane trafficking by the Na/K-ATPase. *J Cell Biol.* 2008;182:1153–1169.

88.   Liu L, Askari A. Beta-subunit of cardiac Na+-K+-ATPase dictates the concentration of the functional enzyme in caveolae. *Am J Physiol, Cell Physiol.* 2006;291:C569-578.

89.   Barbuti A, Scavone A, Mazzocchi N, Terragni B, Baruscotti M, DiFrancesco D. A caveolin-binding domain in the HCN4 channels mediates functional interaction with caveolin proteins. *Journal of Molecular and Cellular Cardiology.* 2012;53:187–195.

90.   Rybin VO, Xu X, Lisanti MP, Steinberg SF. Differential Targeting of β-Adrenergic Receptor Subtypes and Adenylyl Cyclase to Cardiomyocyte Caveolae A MECHANISM TO FUNCTIONALLY REGULATE THE cAMP SIGNALING PATHWAY. *J Biol Chem.* 2000;275:41447–41457.

91.   Fecchi K, Volonte D, Hezel MP, Schmeck K, Galbiati F. Spatial and temporal regulation of GLUT4 translocation by flotillin-1 and caveolin-3 in skeletal muscle cells. *FASEB J.* 2006;20:705–707.

92.   Bossuyt J, Taylor BE, James-Kracke M, Hale CC. Evidence for cardiac sodium-calcium exchanger association with caveolin-3. *FEBS Letters.* 2002;511:113–117.

93.   Vaidyanathan R, Vega AL, Song C, Zhou Q, Tan B, Berger S, Makielski JC, Eckhardt LL. The Interaction of Caveolin 3 Protein with the Potassium Inward Rectifier Channel Kir2.1. *J Biol Chem.* 2013;288:17472–17480.

94.   Dalton DA, Diaz del Castillo L, Kahn ML, Joyner SL, Chatfield JM. Heterologous expression and characterization of soybean cytosolic ascorbate peroxidase. *Arch Biochem Biophys.* 1996;328:1–8.

95.   Lam SS, Martell JD, Kamer KJ, Deerinck TJ, Ellisman MH, Mootha VK, Ting AY. Directed evolution of APEX2 for electron microscopy and proximity labeling. *Nat Methods.* 2015;12:51–54.

96.   Chapman-Smith A, Cronan JE. Molecular biology of biotin attachment to proteins. *J Nutr.* 1999;129:477S-484S.

97.   Rhee H-W, Zou P, Udeshi ND, Martell JD, Mootha VK, Carr SA, Ting AY. Proteomic Mapping of Mitochondria in Living Cells via Spatially Restricted Enzymatic Tagging. *Science.* 2013;339:1328–1331.

98.  Bendayan M. Worth Its Weight in Gold. *Science*. 2001;291:1363–1365.

99.  Lobingier BT, Hüttenhain R, Eichel K, Miller KB, Ting AY, von Zastrow M, Krogan NJ. An Approach to Spatiotemporally Resolve Protein Interaction Networks in Living Cells. *Cell*. 2017;169:350-360.e12.

100.  Maniatis NA, Shinin V, Schraufnagel DE, Okada S, Vogel SM, Malik AB, Minshall RD. Increased pulmonary vascular resistance and defective pulmonary artery filling in caveolin-1-/- mice. *Am J Physiol Lung Cell Mol Physiol*. 2008;294:L865-873.

101.  Cohen AW, Park DS, Woodman SE, Williams TM, Chandra M, Shirani J, Pereira de Souza A, Kitsis RN, Russell RG, Weiss LM, Tang B, Jelicks LA, Factor SM, Shtutin V, Tanowitz HB, Lisanti MP. Caveolin-1 null mice develop cardiac hypertrophy with hyperactivation of p42/44 MAP kinase in cardiac fibroblasts. *American Journal of Physiology-Cell Physiology*. 2003;284:C457–C474.

102.  Couet J, Shengwen L, Okamoto T, Scherer PE, Lisanti MP. Molecular and Cellular Biology of Caveolae: Paradoxes and Plasticities. *Trends in Cardiovascular Medicine*. 1997;7:103–110.

103.  Hsu Y-C, Wang L-F, Chien YW. Nitric oxide in the pathogenesis of diffuse pulmonary fibrosis. *Free Radic Biol Med*. 2007;42:599–607.

104.  Weber KT, Brilla CG, Campbell SE, Zhou G, Matsubara L, Guarda E. Pathologic hypertrophy with fibrosis: the structural basis for myocardial failure. *Blood Press*. 1992;1:75–85.

105.  Yang K-C, Rutledge CA, Mao M, Bakhshi FR, Xie A, Liu H, Bonini MG, Patel HH, Minshall RD, Dudley SC. Caveolin-1 Modulates Cardiac Gap Junction Homeostasis and Arrhythmogenecity by Regulating cSrc Tyrosine Kinase. *Circ Arrhythm Electrophysiol*. 2014;7:701–710.

106.  van Rijen Harold V.M., Eckardt Dominik, Degen Joachim, Theis Martin, Ott Thomas, Willecke Klaus, Jongsma Habo J., Opthof Tobias, de Bakker Jacques M.T. Slow Conduction and Enhanced Anisotropy Increase the Propensity for Ventricular Tachyarrhythmias in Adult Mice With Induced Deletion of Connexin43. *Circulation*. 2004;109:1048–1055.

107.  Woodman SE, Park DS, Cohen AW, Cheung MW-C, Chandra M, Shirani J, Tang B, Jelicks LA, Kitsis RN, Christ GJ, Factor SM, Tanowitz HB, Lisanti MP. Caveolin-3 knock-out mice develop a progressive cardiomyopathy and show hyperactivation of the p42/44 MAPK cascade. *J Biol Chem*. 2002;277:38988–38997.

108. Bryant SM, Kong CHT, Watson JJ, Gadeberg HC, Roth DM, Patel HH, Cannell MB, James AF, Orchard CH. Caveolin-3 KO disrupts t-tubule structure and decreases t-tubular ICa density in mouse ventricular myocytes. *Am J Physiol Heart Circ Physiol*. 2018;315:H1101–H1111.

109. Guo A, Zhang C, Wei S, Chen B, Song L-S. Emerging mechanisms of T-tubule remodelling in heart failure. *Cardiovasc Res*. 2013;98:204–215.

110. Austin ED, Ma L, LeDuc C, Berman Rosenzweig E, Borczuk A, Phillips JA, Palomero T, Sumazin P, Kim HR, Talati MH, West J, Loyd JE, Chung WK. Whole exome sequencing to identify a novel gene (caveolin-1) associated with human pulmonary arterial hypertension. *Circ Cardiovasc Genet*. 2012;5:336–343.

111. Kim CA, Delépine M, Boutet E, El Mourabit H, Le Lay S, Meier M, Nemani M, Bridel E, Leite CC, Bertola DR, Semple RK, O'Rahilly S, Dugail I, Capeau J, Lathrop M, Magré J. Association of a homozygous nonsense caveolin-1 mutation with Berardinelli-Seip congenital lipodystrophy. *J Clin Endocrinol Metab*. 2008;93:1129–1134.

112. Holm H, Gudbjartsson DF, Arnar DO, Thorleifsson G, Thorgeirsson G, Stefansdottir H, Gudjonsson SA, Jonasdottir A, Mathiesen EB, Njølstad I, Nyrnes A, Wilsgaard T, Hald EM, Hveem K, Stoltenberg C, Løchen M-L, Kong A, Thorsteinsdottir U, Stefansson K. Several common variants modulate heart rate, PR interval and QRS duration. *Nat Genet*. 2010;42:117–122.

113. Ellinor PT, Lunetta KL, Albert CM, Glazer NL, Ritchie MD, Smith AV, Arking DE, Müller-Nurasyid M, Krijthe BP, Lubitz SA, Bis JC, Chung MK, Dörr M, Ozaki K, Roberts JD, Smith JG, Pfeufer A, Sinner MF, Lohman K, Ding J, Smith NL, Smith JD, Rienstra M, Rice KM, Van Wagoner DR, Magnani JW, Wakili R, Clauss S, Rotter JI, Steinbeck G, Launer LJ, Davies RW, Borkovich M, Harris TB, Lin H, Völker U, Völzke H, Milan DJ, Hofman A, Boerwinkle E, Chen LY, Soliman EZ, Voight BF, Li G, Chakravarti A, Kubo M, Tedrow UB, Rose LM, Ridker PM, Conen D, Tsunoda T, Furukawa T, Sotoodehnia N, Xu S, Kamatani N, Levy D, Nakamura Y, Parvez B, Mahida S, Furie KL, Rosand J, Muhammad R, Psaty BM, Meitinger T, Perz S, Wichmann H-E, Witteman JCM, Kao WHL, Kathiresan S, Roden DM, Uitterlinden AG, Rivadeneira F, McKnight B, Sjögren M, Newman AB, Liu Y, Gollob MH, Melander O, Tanaka T, Stricker BHC, Felix SB, Alonso A, Darbar D, Barnard J, Chasman DI, Heckbert SR, Benjamin EJ, Gudnason V, Kääb S. Meta-analysis identifies six new susceptibility loci for atrial fibrillation. *Nat Genet*. 2012;44:670–675.

114. Yi S, Liu X, Zhong J, Zhang Y. Role of Caveolin-1 in Atrial Fibrillation as an Anti-Fibrotic Signaling Molecule in Human Atrial Fibroblasts. *PLOS ONE*. 2014;9:e85144.

115. Betz RC, Schoser BGH, Kasper D, Ricker K, Ramírez A, Stein V, Torbergsen T, Lee Y-A, Nöthen MM, Wienker TF, Malin J-P, Propping P, Reis A, Mortier W, Jentsch TJ, Vorgerd M, Kubisch C. Mutations in CAV3 cause mechanical hyperirritability of skeletal muscle in rippling muscle disease. *Nat Genet*. 2001;28:218–219.

116. Carbone I, Bruno C, Sotgia F, Bado M, Broda P, Masetti E, Panella A, Zara F, Bricarelli FD, Cordone G, Lisanti MP, Minetti C. Mutation in the CAV3 gene causes partial caveolin-3 deficiency and hyperCKemia. *Neurology*. 2000;54:1373–1376.

117. Galbiati F, Volonte D, Minetti C, Bregman D, Lisanti M. Limb-girdle muscular dystrophy (LGMD-1C) mutants of caveolin-3 undergo ubiquitination and proteasomal degradation. Treatment with proteasomal inhibitors blocks the dominant negative effect of LGMD-1C mutanta and rescues wild-type caveolin-3. *The Journal of biological chemistry*. 2001;275:37702–11.

118. Traverso M, Gazzerro E, Assereto S, Sotgia F, Biancheri R, Stringara S, Giberti L, Pedemonte M, Wang X, Scapolan S, Pasquini E, Donati MA, Zara F, Lisanti MP, Bruno C, Minetti C. Caveolin-3 T78M and T78K missense mutations lead to different phenotypes in vivo and in vitro. *Lab Invest*. 2008;88:275–283.

119. Minetti C, Bado M, Broda P, Sotgia F, Bruno C, Galbiati F, Volonte D, Lucania G, Pavan A, Bonilla E, Lisanti MP, Cordone G. Impairment of caveolae formation and T-system disorganization in human muscular dystrophy with caveolin-3 deficiency. *Am J Pathol*. 2002;160:265–270.

120. Minetti C, Sotgia F, Bruno C, Scartezzini P, Broda P, Bado M, Masetti E, Mazzocco M, Egeo A, Donati MA, Volonté D, Galbiati F, Cordone G, Bricarelli FD, Lisanti MP, Zara F. Mutations in the caveolin-3 gene cause autosomal dominant limb-girdle muscular dystrophy. *Nature Genetics*. 1998;18:365–368.

121. Galbiati F, Volonté D, Minetti C, Chu JB, Lisanti MP. Phenotypic Behavior of Caveolin-3 Mutations That Cause Autosomal Dominant Limb Girdle Muscular Dystrophy (LGMD-1C) RETENTION OF LGMD-1C CAVEOLIN-3 MUTANTS WITHIN THE GOLGI COMPLEX. *J Biol Chem*. 1999;274:25632–25641.

122. Deng YF, Huang YY, Lu WS, Huang YH, Xian J, Wei HQ, Huang Q. The Caveolin-3 P104L mutation of LGMD-1C leads to disordered glucose

metabolism in muscle cells. *Biochem Biophys Res Commun*. 2017;486:218–223.

123. Cheng J, Valdivia CR, Vaidyanathan R, Balijepalli RC, Ackerman MJ, Makielski JC. Caveolin-3 suppresses late sodium current by inhibiting nNOS-dependent S-nitrosylation of SCN5A. *J Mol Cell Cardiol*. 2013;61:102–110.

124. Gonzalez DR, Treuer A, Sun Q-A, Stamler JS, Hare JM. S-nitrosylation of cardiac ion channels. *J Cardiovasc Pharmacol*. 2009;54:188–195.

125. Sayed N, Liu C, Wu JC. Translation of Human-Induced Pluripotent Stem Cells: From Clinical Trial in a Dish to Precision Medicine. *Journal of the American College of Cardiology*. 2016;67:2161–2176.

126. Cyganek L, Tiburcy M, Sekeres K, Gerstenberg K, Bohnenberger H, Lenz C, Henze S, Stauske M, Salinas G, Zimmermann W-H, Hasenfuss G, Guan K. Deep phenotyping of human induced pluripotent stem cell–derived atrial and ventricular cardiomyocytes. *JCI Insight*. 2018;3.

127. Moretti A, Bellin M, Welling A, Jung CB, Lam JT, Bott-Flügel L, Dorn T, Goedel A, Höhnke C, Hofmann F, Seyfarth M, Sinnecker D, Schömig A, Laugwitz K-L. Patient-Specific Induced Pluripotent Stem-Cell Models for Long-QT Syndrome. *N Engl J Med*. 2010;363:1397–1409.

128. Patmanathan SN, Gnanasegaran N, Lim MN, Husaini R, Fakiruddin KS, Zakaria Z. CRISPR/Cas9 in Stem Cell Research: Current Application and Future Perspective. *Curr Stem Cell Res Ther*. 2018;13:632–644.

129. Sterneckert JL, Reinhardt P, Schöler HR. Investigating human disease using stem cell models. *Nature Reviews Genetics*. 2014;15:625–639.

130. Maeder ML, Gersbach CA. Genome-editing Technologies for Gene and Cell Therapy. *Molecular Therapy*. 2016;24:430–446.

131. Harkness L, Chen X, Gillard M, Gray PP, Davies AM. Media composition modulates human embryonic stem cell morphology and may influence preferential lineage differentiation potential. *PLOS ONE*. 2019;14:e0213678.

132. Rana P, Anson B, Engle S, Will Y. Characterization of Human-Induced Pluripotent Stem Cell–Derived Cardiomyocytes: Bioenergetics and Utilization in Safety Screening. *Toxicol Sci*. 2012;130:117–131.

133. Wagner E, Brandenburg S, Kohl T, Lehnart SE. Analysis of Tubular Membrane Networks in Cardiac Myocytes from Atria and Ventricles. *J Vis Exp*. 2014;

134. Oliveira GS, Santos AR. Using the Gene Ontology tool to produce de novo protein-protein interaction networks with IS_A relationship. *Genet Mol Res.* 2016;15.

135. Keshavarz M, Skill M, Hollenhorst MI, Maxeiner S, Walecki M, Pfeil U, Kummer W, Krasteva-Christ G. Caveolin-3 differentially orchestrates cholinergic and serotonergic constriction of murine airways. *Sci Rep.* 2018;8:1–18.

136. Matsuda C, Hayashi YK, Ogawa M, Aoki M, Murayama K, Nishino I, Nonaka I, Arahata K, Jr RHB. The sarcolemmal proteins dysferlin and caveolin-3 interact in skeletal muscle. *Hum Mol Genet.* 2001;10:1761–1766.

137. Tusher VG, Tibshirani R, Chu G. Significance analysis of microarrays applied to the ionizing radiation response. *Proc Natl Acad Sci USA.* 2001;98:5116–5121.

138. Erdmann J, Thöming JG, Pohl S, Pich A, Lenz C, Häussler S. The Core Proteome of Biofilm-Grown Clinical Pseudomonas aeruginosa Isolates. *Cells.* 2019;8:1129.

139. Parton RG, Collins BM. Unraveling the architecture of caveolae. *Proc Natl Acad Sci USA.* 2016;113:14170–14172.

140. Bastiani M, Parton RG. Caveolae at a glance. *J Cell Sci.* 2010;123:3831–3836.

141. Hentze MW, Muckenthaler MU, Andrews NC. Balancing acts: molecular control of mammalian iron metabolism. *Cell.* 2004;117:285–297.

142. Szklarczyk D, Gable AL, Lyon D, Junge A, Wyder S, Huerta-Cepas J, Simonovic M, Doncheva NT, Morris JH, Bork P, Jensen LJ, Mering C von. STRING v11: protein–protein association networks with increased coverage, supporting functional discovery in genome-wide experimental datasets. *Nucleic Acids Research.* 2019;47:D607–D613.

143. Fujimoto T, Kogo H, Nomura R, Une T. Isoforms of caveolin-1 and caveolar structure. *J Cell Sci.* 2000;113 Pt 19:3509–3517.

144. Roux KJ, Kim DI, Raida M, Burke B. A promiscuous biotin ligase fusion protein identifies proximal and interacting proteins in mammalian cells. *J Cell Biol.* 2012;196:801–810.

145. Hung V, Zou P, Rhee H-W, Udeshi ND, Cracan V, Svinkina T, Carr SA, Mootha VK, Ting AY. Proteomic Mapping of the Human Mitochondrial Intermembrane Space in Live Cells via Ratiometric APEX Tagging. *Molecular Cell.* 2014;55:332–341.

146. Kim DI, Birendra KC, Zhu W, Motamedchaboki K, Doye V, Roux KJ. Probing nuclear pore complex architecture with proximity-dependent biotinylation. *Proc Natl Acad Sci USA*. 2014;111:E2453-2461.

147. Roux KJ, Kim DI, Burke B, May DG. BioID: A Screen for Protein-Protein Interactions. *Curr Protoc Protein Sci*. 2018;91:19.23.1-19.23.15.

148. Li T, Zhang X, Jiang K, Liu J, Liu Z. Dural effects of oxidative stress on cardiomyogenesis via Gata4 transcription and protein ubiquitination. *Cell Death Dis*. 2018;9:1–12.

149. Fernandez-Caggiano M, Schröder E, Cho H-J, Burgoyne J, Barallobre-Barreiro J, Mayr M, Eaton P. Oxidant-induced Interprotein Disulfide Formation in Cardiac Protein DJ-1 Occurs via an Interaction with Peroxiredoxin 2. *J Biol Chem*. 2016;291:10399–10410.

150. Jiang H, White EJ, Ríos-Vicil CI, Xu J, Gomez-Manzano C, Fueyo J. Human Adenovirus Type 5 Induces Cell Lysis through Autophagy and Autophagy-Triggered Caspase Activity ▽. *J Virol*. 2011;85:4720–4729.

151. Jóhannsson E, Nagelhus EA, McCullagh KJ, Sejersted OM, Blackstad TW, Bonen A, Ottersen OP. Cellular and subcellular expression of the monocarboxylate transporter MCT1 in rat heart. A high-resolution immunogold analysis. *Circ Res*. 1997;80:400–407.

152. Soeiro M de NC, Mota RA, Batista D da GJ, Meirelles M de NL. Endocytic Pathway in Mouse Cardiac Cells. *Cell Struct Funct*. 2002;27:469–478.

153. Xu W, Barrientos T, Andrews NC. Iron and copper in mitochondrial diseases. *Cell Metab*. 2013;17:319–328.

154. Abbaspour N, Hurrell R, Kelishadi R. Review on iron and its importance for human health. *J Res Med Sci*. 2014;19:164–174.

155. Zhang C. Essential functions of iron-requiring proteins in DNA replication, repair and cell cycle control. *Protein Cell*. 2014;5:750–760.

156. Xu W, Barrientos T, Mao L, Rockman HA, Sauve AA, Andrews NC. Lethal Cardiomyopathy in Mice Lacking Transferrin Receptor in the Heart. *Cell Rep*. 2015;13:533–545.

157. Gulati V, Harikrishnan P, Palaniswamy C, Aronow WS, Jain D, Frishman WH. Cardiac involvement in hemochromatosis. *Cardiol Rev*. 2014;22:56–68.

158. Hoes MF, Grote Beverborg N, Kijlstra JD, Kuipers J, Swinkels DW, Giepmans BNG, Rodenburg RJ, van Veldhuisen DJ, de Boer RA, van der Meer P. Iron deficiency impairs contractility of human cardiomyocytes

through decreased mitochondrial function. *Eur J Heart Fail*. 2018;20:910–919.

159. Garcia CK, Goldstein JL, Pathak RK, Anderson RG, Brown MS. Molecular characterization of a membrane transporter for lactate, pyruvate, and other monocarboxylates: implications for the Cori cycle. *Cell*. 1994;76:865–873.

160. Levy B, Mansart A, Montemont C, Gibot S, Mallie J-P, Regnault V, Lecompte T, Lacolley P. Myocardial lactate deprivation is associated with decreased cardiovascular performance, decreased myocardial energetics, and early death in endotoxic shock. *Intensive Care Med*. 2007;33:495–502.

161. van der Vusse GJ, de Groot MJ. Interrelationship between lactate and cardiac fatty acid metabolism. *Mol Cell Biochem*. 1992;116:11–17.

162. Jóhannsson E, Lunde PK, Heddle C, Sjaastad I, Thomas MJ, Bergersen L, Halestrap AP, Blackstad TW, Ottersen OP, Sejersted OM. Upregulation of the Cardiac Monocarboxylate Transporter MCT1 in a Rat Model of Congestive Heart Failure. *Circulation*. 2001;104:729–734.

163. Vandenberg JI, Metcalfe JC, Grace AA. Mechanisms of pHi recovery after global ischemia in the perfused heart. *Circ Res*. 1993;72:993–1003.

164. Xu KY, Zweier JL, Becker LC. Functional coupling between glycolysis and sarcoplasmic reticulum Ca2+ transport. *Circ Res*. 1995;77:88–97.

165. Murphy E, Steenbergen C. Ion Transport and Energetics During Cell Death and Protection. *Physiology (Bethesda)*. 2008;23:115–123.

166. Doherty JR, Yang C, Scott KEN, Cameron MD, Fallahi M, Li W, Hall MA, Amelio AL, Mishra JK, Li F, Tortosa M, Genau HM, Rounbehler RJ, Lu Y, Dang ChiV, Kumar KG, Butler AA, Bannister TD, Hooper AT, Unsal-Kacmaz K, Roush WR, Cleveland JL. Blocking Lactate Export by Inhibiting the Myc Target MCT1 Disables Glycolysis and Glutathione Synthesis. *Cancer Res*. 2014;74:908–920.

167. Reijneveld JC, Ginjaar IB, Frankhuizen WS, Notermans NC. CAV3 gene mutation analysis in patients with idiopathic hyper-CK-emia. *Muscle Nerve*. 2006;34:656–658.

168. Tsutsumi YM, Kawaraguchi Y, Horikawa YT, Niesman IR, Kidd MW, Chin-Lee B, Head BP, Patel PM, Roth DM, Patel HH. Role for Caveolin-3 and Glucose Transporter 4 in Isoflurane-induced Delayed Cardiac Protection. *Anesthesiology*. 2010;112:1136–1145.

169. Huang S, Czech MP. The GLUT4 Glucose Transporter. *Cell Metabolism.* 2007;5:237–252.

170. Abel ED, Kaulbach HC, Tian R, Hopkins JCA, Duffy J, Doetschman T, Minnemann T, Boers M-E, Hadro E, Oberste-Berghaus C, Quist W, Lowell BB, Ingwall JS, Kahn BB. Cardiac hypertrophy with preserved contractile function after selective deletion of GLUT4 from the heart. *J Clin Invest.* 1999;104:1703–1714.

171. Fecchi K, Volonte D, Hezel MP, Schmeck K, Galbiati F. Spatial and temporal regulation of GLUT4 translocation by flotillin-1 and caveolin-3 in skeletal muscle cells. *FASEB J.* 2006;20:705–707.

172. Ralston E, Ploug T. Caveolin-3 is associated with the T-tubules of mature skeletal muscle fibers. *Exp Cell Res.* 1999;246:510–515.

173. Capozza F, Combs TP, Cohen AW, Cho Y-R, Park S-Y, Schubert W, Williams TM, Brasaemle DL, Jelicks LA, Scherer PE, Kim JK, Lisanti MP. Caveolin-3 knockout mice show increased adiposity and whole body insulin resistance, with ligand-induced insulin receptor instability in skeletal muscle. *Am J Physiol, Cell Physiol.* 2005;288:C1317-1331.

174. Kogo H, Aiba T, Fujimoto T. Cell Type-specific Occurrence of Caveolin-1α and -1β in the Lung Caused by Expression of Distinct mRNAs. *J Biol Chem.* 2004;279:25574–25581.

175. Gottlieb-Abraham E, Shvartsman DE, Donaldson JC, Ehrlich M, Gutman O, Martin GS, Henis YI. Src-mediated caveolin-1 phosphorylation affects the targeting of active Src to specific membrane sites. *Mol Biol Cell.* 2013;24:3881–3895.

176. Li S, Seitz R, Lisanti MP. Phosphorylation of caveolin by src tyrosine kinases. The alpha-isoform of caveolin is selectively phosphorylated by v-Src in vivo. *J Biol Chem.* 1996;271:3863–3868.

177. Zimnicka AM, Husain YS, Shajahan AN, Sverdlov M, Chaga O, Chen Z, Toth PT, Klomp J, Karginov AV, Tiruppathi C, Malik AB, Minshall RD. Src-dependent phosphorylation of caveolin-1 Tyr-14 promotes swelling and release of caveolae. *Mol Biol Cell.* 2016;27:2090–2106.

178. Cao H, Courchesne WE, Mastick CC. A phosphotyrosine-dependent protein interaction screen reveals a role for phosphorylation of caveolin-1 on tyrosine 14: recruitment of C-terminal Src kinase. *J Biol Chem.* 2002;277:8771–8774.

179. Lee H, Volonte D, Galbiati F, Iyengar P, Lublin DM, Bregman DB, Wilson MT, Campos-Gonzalez R, Bouzahzah B, Pestell RG, Scherer PE, Lisanti

MP. Constitutive and growth factor-regulated phosphorylation of caveolin-1 occurs at the same site (Tyr-14) in vivo: identification of a c-Src/Cav-1/Grb7 signaling cassette. *Mol Endocrinol.* 2000;14:1750–1775.

180. Mathai JC, Mori S, Smith BL, Preston GM, Mohandas N, Collins M, van Zijl PC, Zeidel ML, Agre P. Functional analysis of aquaporin-1 deficient red cells. The Colton-null phenotype. *J Biol Chem.* 1996;271:1309–1313.

181. Rutkovskiy A, Bliksøen M, Hillestad V, Amin M, Czibik G, Valen G, Vaage J, Amiry-Moghaddam M, Stensløkken K-O. Aquaporin-1 in cardiac endothelial cells is downregulated in ischemia, hypoxia and cardioplegia. *Journal of molecular and cellular cardiology.* 2012;56.

182. Gao C, Li R, Huan J, Li W. Caveolin-1 siRNA increases the pulmonary microvascular and alveolar epithelial permeability in rats. *J Trauma.* 2011;70:210–219.

183. Butler TL, Au CG, Yang B, Egan JR, Tan YM, Hardeman EC, North KN, Verkman AS, Winlaw DS. Cardiac aquaporin expression in humans, rats, and mice. *Am J Physiol Heart Circ Physiol.* 2006;291:H705-713.

184. Verkerk AO, Lodder EM, Wilders R. Aquaporin Channels in the Heart-Physiology and Pathophysiology. *Int J Mol Sci.* 2019;20.

185. Fernández-Jiménez R, Galán-Arriola C, Sánchez-González J, Agüero J, López-Martín GJ, Gomez-Talavera S, Garcia-Prieto J, Benn A, Molina-Iracheta A, Barreiro-Pérez M, Martin-García A, García-Lunar I, Pizarro G, Sanz J, Sánchez PL, Fuster V, Ibanez B. Effect of Ischemia Duration and Protective Interventions on the Temporal Dynamics of Tissue Composition After Myocardial Infarction. *Circ Res.* 2017;121:439–450.

186. Kloner RA, Jennings RB. Consequences of brief ischemia: stunning, preconditioning, and their clinical implications: part 2. *Circulation.* 2001;104:3158–3167.

www.ingramcontent.com/pod-product-compliance
Lightning Source LLC
LaVergne TN
LVHW041656190726
843493LV00007B/1819